FUZZY LOGIC
for Embedded Systems Applications

FUZZY LOGIC
for Embedded Systems Applications

by

Ahmad M. Ibrahim, Ph.D.
Senior Member, IEEE

ELSEVIER

AMSTERDAM • BOSTON • HEIDELBERG • LONDON
NEW YORK • OXFORD • PARIS • SAN DIEGO
SAN FRANCISCO • SINGAPORE • SYDNEY • TOKYO

Newnes

Newnes is an imprint of Elsevier Science.

Library of Congress Cataloging-in-Publication Data

ISBN: 0-7506-7605-1

British Library Cataloging-in-Publication Data
A catalogue record for this book is available from the British Library.

The publisher offers special discounts on bulk orders of this book.
For information, please contact:

Manager of Special Sales
Elsevier Science
200 Wheeler Road
Burlington, MA 01803
Tel: 781-313-4700
Fax: 781-313-4880

For information on all Newnes publications available, contact our World Wide Web home page at: http://www.newnespress.com

10 9 8 7 6 5 4 3 2 1

Printed in the United States of America

Contents

Preface

Fuzzy logic and its applications are now well-established and arguments for and against it have reached a steady state. There is an overwhelming volume of literature on the topic making it a difficult task for a practicing engineer, beginner researcher, or an advanced student to grasp the topic and then apply the acquired knowledge with only a small investment of time and money. This book is intended to present fuzzy logic and its applications for embedded systems succinctly yet comprehensively, with a self-contained, simple, readable approach. Simplicity here means the omission of extraneous sentences and phrases, and the exclusion of non-applied mathematical and research-oriented details. It is intended for the intelligent reader with an alert mind. The approach, the organization, and the presentation of this book are also hoped to enhance the accessibility to existing knowledge beyond its contents. An extensive bibliography not only of printed material but also of annotated Web links is provided at the end of each chapter.

The book is divided into nine chapters in addition to a set of quizzes, an appendix, a list of symbols and acronyms, and a glossary. Chapter 1 gives an overview of embedded systems and their implementation techniques. The chapter introduces the wide scope of embedded systems. The relationship between fuzzy logic and embedded systems is also outlined.

Chapters 2 and 3 introduce succinctly the concepts of fuzzy sets, fuzzy operations, and fuzzy relations with illustrative examples. The discussion is geared toward what would be needed in order to design fuzzy embedded systems.

In Chapter 4, embedded fuzzy logic applications are introduced with simplified case studies. Contrasting fuzzy logic control with conventional control is emphasized. It is hoped that by the end of this chapter that the reader would be able to apply fuzzy logic to the design of an embedded system of interest. The reader may wish to consult Chapter 8, Fuzzy Software Tools, select a few tools and start experimenting with fuzzy systems design and simulation.

A critique of fuzzy logic is presented in Chapter 5, a topic that is not commonly discussed in the engineering literature on fuzzy logic. It is, however, important for engineers to understand the limitations and implications of any methodology they intend to use. No tool is suitable for everything; it must be used skillfully within its bounds of applicability. Moreover, most engineers are interested in intellectual discussions relevant to the topic at hand as long as the ideas are communicated properly and to the point. The discussion in this chapter is brief, but with an extensive bibliography provided for those who may like to further research any of the points discussed.

Chapter 6 presents the fundamentals of artificial neural networks, which are closely related to fuzzy logic but with fundamentally different concepts. The chapter introduces basic structures and learning algorithms of neural networks.

Hardware realizations are outlined in Chapter 8, which discusses both analog and digital implementations. Chapter 9 provides the reader with an opportunity to gain hands-on experience with a minimum investment of time and money. The chapter gives an overview of numerous software tools for fuzzy logic systems, neural networks, and neuro-fuzzy systems. The software reviewed includes C, C++, and Java source codes, in addition to M-files for MATLAB. Design and analysis software tools with graphical user interfaces are also discussed, as well as software that is meant to demonstrate fundamental concepts. All software tools, or working demos of them, are available for downloading through the Web along with documentation and examples in most cases.

Since most of the ideas presented are now well established there are no references given within the chapter text. However, further reading and expansion on the information presented is facilitated by the selected bibliography provided at the end of each chapter. The Web resources selected are those available at no charge, relevant to the topics discussed, and expected to be reliable.

The idea of citing references from the Web does not yet hold wide acceptance in some academic circles. The major argument against the idea is the questionable reliability of the content and its sustained availability. It should be remembered that content does not become reliable just because it is printed on paper, and unreliable when it is posted freely on the Web. Nevertheless, many of the resources cited as accessible through the Web are documents that were published on paper as well, and the Web simply enhances their availability. The use of the Web is essential for some of the resources such as interactive demos. It should also be remembered that books are known to go out of print and many may not be physically accessible with ease. Of course, an available, well written book constitutes a better and more convenient source of knowledge. Accessibility through the Web would be the next best thing, but it requires guidance, as provided here, to make the most out of it.

The set of quizzes with answers provided is meant to help the reader ponder about the subjects introduced without being side-tracked from the final goal that could be better reached by using the resources of chapter 9; thus learning through practice.

The appendix introduces the fundamental concepts of Genetic Algorithms (GAs). It is an optimization technique sometimes used in conjunction with the design of fuzzy and neural systems. A classified, annotated Web bibliography is provided for the reader who may wish to study the topic further.

A list of symbols and acronyms is provided along with the circuit symbols of MOSFETs that appear in literature on fuzzy logic embedded implementations to help the reader avoid possible confusion when consulting the resources.

A glossary of terms related to embedded systems, fuzzy logic and neural networks is provided. Although it refers to terms used in the various chapters of the book, it also introduces some terms to expand the coverage and provoke further interest in the general topic.

If an instructor is to use this book as a teaching resource, numerous application-oriented exercises and mini-projects can be assigned to the students if the instructor wishes to do so. For example:

- The quizzes provided could be the basis for class discussion on the fundamentals of fuzzy logic and neural networks.

- An item from the numerous Web resources provided could be selected and students asked to research it further and expand on the short review given.

- Students could be asked to research entries from the glossary further.

- Students could be asked to run a selected demo simulation or Java applet from the resources provided in Chapter 9, observe its action, document their observations, and relate it to the theory.

- Students could be assigned a system of interest such as a refrigerator, vacuum cleaner, auto-focus, level control, speed control, etc. They could identify the inputs and outputs, design a linguistic model, then simulate it using one of the tools discussed.

- Students could be also asked to produce their own source code and executable file to solve one of the above problems.

- A topic from Chapter 5, Fuzzy Logic Critique, could be selected for further research and discussion.

It is hoped that through this book the reader will:

- Gain an understanding of the wide range of embedded systems and their future trend.

- Be able to use fuzzy sets and fuzzy logic algebra.

- Recognize when and why it would be advantageous to use fuzzy logic.

- Understand the fundamentals of neural networks and recognize when it is advantageous to use them.

- Gain familiarity with the hardware implementation approaches of fuzzy logic for embedded systems applications.

- Be able to experiment with the design of fuzzy systems for embedded systems applications.

- Be able to pursue further details of a topic within fuzzy logic embedded systems applications with relative ease.

I would like to express my gratitude to Dr. Alexander Berezin, McMaster University, Hamilton, Ontario, Canada for his encouragement and useful discussions. I would also like to thank Dr. Lawrence Hulsman, University of Arizona, Tucson, USA for his continuous encouragement and inspiration. The comments provided by Mahmoud Gadala, Senior Engineer, Pratt & Whitney Canada Inc, Montreal, Quebec, Canada are highly appreciated. I would also like to thank Dr. Ahmed Hussein, University of Northern British Columbia, Prince George, British Columbia, Canada, Dr. Erol Inelmen, Bogazici University, Istanbul, Turkey, Prof. Predrag Pesikan, DeVry, Mississauga, Ontario, Canada, and Dr. Aleksander Malinowski, Bradley University, Peoria, Illinois, USA; their interest and input are particularly valued.

About the Author

Dr. Ibrahim is a senior member of the Institute of Electrical and Electronics Engineers (IEEE), a member of the Association of Professional Engineers of Ontario (APEO), the Material Research Society (MRS), the American Association of Engineering Education (ASEE), and the International Banknote Society (IBS). He lectured in the area of electronics on three continents to a diverse population of students and presented seminars and workshops to practicing engineers.

He has a wide range of publications including papers in refereed journals and conferences. Dr. Ibrahim has organized and chaired sessions on *Current Trends in Electronics Education* during the IEEE Conferences IECON'01, Denver, Colorado, USA, and IECON'02, Sevilla, Andalucia, Spain. He was a guest editor for the 2003 Special Issue of the International Journal of Engineering Education on *Current Trends in Electronics Education.* He was a member of the International Advisory Committee of the First and Second International Conferences on Distanc Learning held at the Belarussian State University of Informatics and Radioelectroi s, Minsk, Belarus in the years 2001 and 2002. Dr. Ibrahim has a B.Sc. (EE) degree from Ain Shams University, Cairo, Egypt, and a Masters (Eng. Phys.) and Ph.D. (EE) degrees from McMaster University, Hamilton, Ontario, Canada. At present he is a senior professor at DeVry, Mississauga, Ontario, Canada.

What's on the CD-ROM?

Included on the accompanying CD-ROM:

- A fully searchable eBook version of the text in Adobe pdf format
- Links to numerous useful fuzzy-related web sites
- Additional informative resources and documents

Access the Read_Me file for more details on the CD-ROM contents.

CHAPTER 1

Embedded Systems: An Overview

1.1 Definition and Examples

A system whose principal function is not computational but is controlled by a computational system embedded within it is referred to as a computational embedded system, which is usually shortened to *embedded system.*

The word *embedded* implies that the computational system lies within an overall system. The user of the overall system may not even be aware of the computational system's existence. The phrase "computational system" refers to computers, microprocessors, microcontrollers, DSP chips, or custom-made hardware along with the software that may be associated with them. Desktop or laptop computers, although they contain microprocessors, are not considered embedded systems since their principal function is computational. If such a computer is customized and built permanently into an identifiable system with the sole purpose of controlling that particular system, then one can refer to it as part of an embedded system.

Embedded systems are used in a wide range of applications. They can be loosely categorized as:

- **Consumer Electronics** including digital cameras, televisions, cell phones, and camcorders.

- **Home Appliances** including microwave ovens, rice cookers, thermostats, washing machines, and drying machines.

- **Business and Office Equipment** including alarm systems, card readers, product scanners, cash registers, fax machines, and copiers.

- **Transportation Systems** including automobiles (transmission control, fuel injection, antilock brakes, and cruise control), train systems, and avionic systems.

- **Factory Control** including machine control, instrumentation, and robotics.

- **Medical Systems** including life-support systems, testing systems, and diagnostic systems.

The list of embedded system applications is growing continuously and is almost endless.

1.2 Further Features of Embedded Systems

Embedded systems share common features that include the following:

- **Hardware and software**

  An embedded computing system typically employs both software and hardware components. The emphasis in the design varies depending on the constraints of the design specifications. The capabilities of integrated circuit (IC) chips have been following the so-called Moore's law, which predicts that IC technology will double its performance every 18 months. Figure 1.1 gives an illustration of Moore's Law, showing the increase in the number of transistors in processors over time.

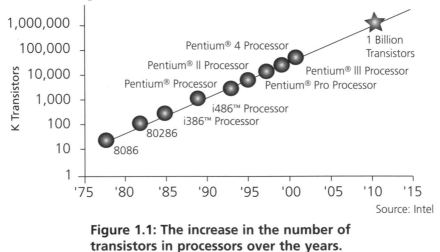

Source: Intel

**Figure 1.1: The increase in the number of
transistors in processors over the years.**

There is, of course, a physical limitation on the dimensions of a transistor which, when reached, means different physical laws become applicable. This may lead to practical application of quantum computing and a new meaning to embedded systems. It appears that the amount of software used per consumer product is doubling roughly every two years. The trend appears to be continuing and the number of familiar and new consumer products using all the capabilities of IC technology is increasing as prices become suitable for consumers. One has to be careful, however, when talking about the *amount* of software. The number of lines of code and their rate of execution would not be enough to evaluate the performance of a system independent of what a code instruction can do.

Manufacturers appear to depend on software to improve the user interface of products. They also depend more and more on software in introducing the product diversity needed in the market while maintaining the minimum hardware variety to ensure economical mass production. Some of the features of embedded computing systems include:

- **Tightly-Constrained Specifications**

  Embedded computing systems typically have particularly tight design specifications. The physical dimensions of the system usually need to be small, requiring the system to fit on a single chip. The cost is required to be low as well. In most cases, the system must continually react to changes in its environment and to process the data fast enough as the changes occur (real-time operation).

- **Reliability**

  Bugs in embedded computing systems cannot be tolerated, as opposed to personal computing where software bugs are mere inconveniences. Rigorous systematic testing and commissioning are required. Implementing an embedded system with hardware or software problems could lead to serious consequences. For example, one significant software error in a life-support system or a transportation system could be fatal.

  The need for assured reliability becomes increasingly demanding considering the fact that numerous products with embedded systems have a relatively short life cycle. They tend to become obsolete due to technological changes.

- **Networked Processors**

  Some embedded computing systems may require the use of more than one processor, each to control one subsystem. These processors need to be networked to share data and to meet the needs of the overall system. For example, an automobile has subsystems that require electronic control. These include fuel injection and ignition, automatic transmissions, air-conditioning systems, shock absorbers, anti-lock braking systems, and air bags. A network of inexpensive processors could lead to a cheaper and simpler system implementation than a single processor performing all the tasks.

- **Internet-enabled**

  The increased popularity of the internet and multimedia has created a demand for embedded consumer product networking. It is predicted that the traditionally separated worlds of entertainment (audio/video equipment), communication (telephone and e-mail), and computing will become inseparable. All, in addition to home appliances (washing machines and refrigerators), would need to be linked to the internet or its successor. Table 1.1 gives a summary of major applications and initiatives of embedded internet application areas.

Table 1.1: Summary of major applications and initiatives of embedded inherent application areas.

Category	Examples of Commercial Applications
Home Appliances and Consumer Electronics	Equipment diagnostics and service, Web-based content distribution
Office Systems	Copier, printer, and fax machine configuration and system management
Factory Control	Machine controller configuration and diagnostics
Medical Systems	Diagnosis, treatment, and data communications
Communications	Switch and network router configuration and management PDAs, cellphones, and pagers for Web-based content distribution
Automotive Systems	GPS and traffic reports Maintenance and diagnostics Entertainment

1.3 Design Metrics

Design metrics are measurable features of the implemented systems. They are useful in characterizing and comparing various implementations of embedded systems. They include:

- **Safety**

  This feature overrides any other. A system that is not safe is not useful. The system needs to be guaranteed that it causes no harm to the individuals using it, the environment, or other systems it may come in contact with.

- **Cost**

  The overall cost includes the cost of designing, testing the prototype, and the cost of manufacturing each unit.

- **Time-to-market**

  This is the time it takes to develop a system and make it available on the market. It includes the time needed to design and test a prototype and the manufacturing time. Time-to-market needs to be much shorter than the lifetime of the product in the market before consumers consider it to be obsolete. This time will have a direct impact on the expected revenues from marketing the product.

■ **Size**

In numerous embedded systems the physical size and weight of the system are important. They are related in many cases to the hardware and software used. Software size is measured in bytes and number of instructions of the code used. The number of gates or transistors are used to measure the size of the hardware.

■ **Performance**

Performance of an embedded system typically refers to the time the system takes to perform key tasks. This would include the response time, i.e. the time between the start of execution of a task and its finish, and the number of tasks performed per unit time. Sometimes the performance of a processor is cited in instructions per unit time; one has to be careful when comparing processors on such a basis to consider what an instruction can do.

■ **Power Consumption**

Power consumption is an important feature. The power consumed by the embedded system determines the lifetime of the system's battery, the cooling requirements of the hardware, and reliability. Figure 1.2 shows that dissipated power has increased over the years as the number of integrated transistors has increased.

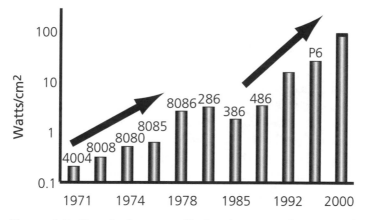

Figure 1.2: Trend of power dissipation over the years. (Source: Intel)

■ **Design Modification Flexibility**

This feature refers to the level of ease with which the functionality of the system can be modified with minimum cost to produce a new prototype. Well-documented systems enable designers, even those who were not involved in the initial design, to maintain and modify a given system economically.

- **Bench Marking**

  The EDN Embedded Microprocessor Benchmark Consortium (EEMBC, pronounced "embassy") was formed in 1997 to develop performance benchmarks for processors for embedded applications. EEMBC's benchmarks are meant to address real-world applications in areas such as automotive/industrial/consumer, networking, telecommunications, and office automation. This would provide designers with a certified method of comparing and judging performance.

  It is important to observe that the metrics are typically correlated, either positively or negatively. Challenges occur and hence there is a need to compromise when the correlation is negative—i.e., improving one feature often leads to degrading another.

1.4 Implementation

Embedded systems implementation relates to three general sets of technologies; processor, hardware and software technologies; IC chip technologies; and design and test technologies. The following sections give a short overview of these categories.

1.4.1 Processor Technologies

General-Purpose Processors

A general-purpose processor, or a microprocessor, is a predesigned programmable digital system that can be used for numerous diverse applications. It is used in embedded systems by programming its memory to execute the required operations.

Such an approach to embedded system design has the advantage of design modification flexibility through program modification. It also has the advantage of a lower prototype development cost and hence a lower cost for the production of a small number of units. In addition, the development time is relatively short. However, the cost is higher than custom-designed processors if the number of units required to be manufactured is large. The size and power consumption could be larger than necessary because of the existence of unneeded processor hardware.

Several software tools are typically available for programming a chosen general-purpose processor. The existence of an Integrated Development Environment (IDE) simplifies the design process further since it enables writing of the source code, its compilation, and linking it into an executable file on the intended microprocessor, all from within a single application

A microcontroller is a special type of microprocessor, designed for embedded applications. It usually features on-chip program and data memory to enable single-chip design solutions.

Limited-Purpose Processors

Limited purpose processors include single-purpose and special-purpose processors. A single-purpose processor is a digital circuit designed to execute one particular program. Other terms used include coprocessor, accelerator, and peripheral. A special-purpose processor is a digital circuit that can execute numerous programs, but its design is optimized for a particular application area, for example digital signal processing (DSP).

The performance of such processors is typically faster than that of general-purpose processors. Also, they are smaller and their power consumption is lower. However, they are less flexible. The unit cost for small quantities and the nonrecurring engineering costs are also higher.

1.4.2 Integrated Circuit Technologies

Processors are implemented using integrated circuit technologies. Integrated circuits (ICs) are sometimes referred to as *chips*. They are typically classified based on the number of transistors with which the chip is built:

SSI (Small-Scale Integration): up to 100 per chip; for example, logic gate chips.

MSI (Medium-Scale Integration): from 100 to 3000; for example, flip-flops and counters chips.

LSI (Large-Scale Integration): from 3000 to 100,000; for example, peripheral interface chips.

VLSI (Very Large-Scale Integration): from 100,000 to 1,000,000; for example, microprocessors, and memory chips.

ULSI (Ultra Large-Scale Integration): more than 1 million.
The line between VLSI and ULSI is vague, and the term ULSI is most commonly used in Japanese literature.

Silicon (Si), and gallium arsenide (GaAs), are semiconductor materials used for IC fabrication. The physical properties of GaAs lead to inherently faster devices, but it is more expensive than Si.

Two technologies are used for silicon-based ICs: Bipolar Junction Transistors (BJT), and Metal-Oxide-Semiconductor Field Effect Transistors (MOSFET). Bipolar technology leads to device families such as TTL (transistor-transistor logic) and I²L (integrated injection logic). The bipolar technology offers high speed and high current drive, but leads to higher power dissipation and lower circuit complexity.

The MOSFET technology offers low power dissipation, a very high level of integration, and better electrical characteristics. Circuit realization techniques may utilize:

- Enhancement type n-channel (EnMOS)

- Both enhancement and depletion type n-channel (EDnMOS)

- Both p-channel and n-channel enhancement MOSFET, which is known as CMOS (complementary metal-oxide semiconductor). CMOS-based implementation is becoming the standard process for all but the highest speed devices. It offers lower power dissipation but the circuit complexity for realization is lower compared to those which may be realized with EDnMOS.

- BiCMOS devices are based on both bipolar and CMOS technologies. They give the best of the two technologies, but lead to increased cost and fabrication complexity.

An IC is fabricated by creating several layers of different characteristics in a Si wafer. Such layers can be created by depositing photosensitive chemicals on the surface of a silicon wafer and shining light through masks to change the properties of the chemicals. An etching process follows to remove parts of the chemicals as required to expose the surface according to a given pattern and make the required changes in the properties as needed. The layers are thus built using an appropriately designed set of masks (referred to as a *layout*). Figure 1.3 shows current and future CMOS structures.

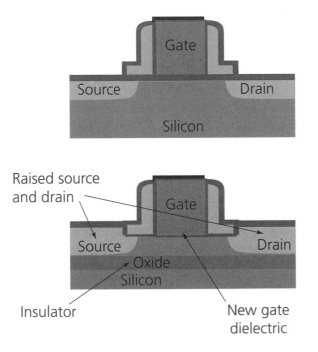

Figure 1.3: Intel CMOS current and future structure. (Source: Intel)

The top illustration of Figure 1.3 shows the basic structure of a CMOS transistor currently in use (year 2003). All CMOS circuits are built with transistors such as this, and connected to each other according to the circuit design. The Pentium® 4 processor has 42 million such transistors integrated on a piece of silicon as small as a fingernail. The lower part of Figure 1.3 illustrates the structure of the new TeraHertz transistor that is expected to be in use within a few years. Its goal is to maintain the level of power consumption of present day microprocessors, although the future ones will be built with many more transistors.

Technology generation has been defined by the feature size, λ, which is the smallest feature on an IC—for example, the length of the transistor channel. Current feature size is in the submicron range. Figure 1.4 shows the trend in feature size over the years.

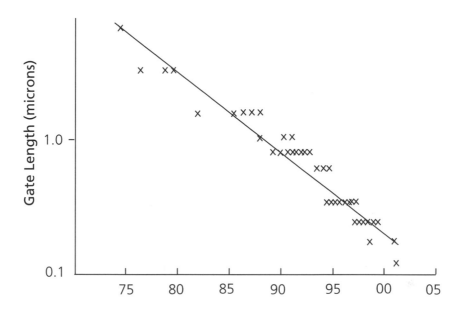

Figure 1.4: Feature size of Intel microprocessors over the years. (Source: Intel)

There are numerous design styles for IC implementation of given logic functions or algorithms, each with its merits and demerits. These include:

Semicustom ASIC
In an application-specific IC, (ASIC), the chip is partially finished in a standard form, and then it is completed as the need arises based on the design to be implemented. ASICs include FPGAs, PLDs, and CPLDs.

Field Programmable Device, FPD

FPD is a general term that refers to any type of integrated circuit used for implementing hardware by user programming. These devices are also referred to as Programmable Logic Devices, PLDs. They are similar in principle to the EPROM (Electrically Programmable Read Only Memory), but they have many more potential applications. SPLD refers to simple PLD, and CPLD refers to complex PLD. A CPLD consists of several interconnected programmable SPLD-like blocks on a single chip. CPLDs feature logic resources with a large number of inputs. A Field Programmable Gate Array (FPGA), is an FPD with a general structure that allows very high logic capacity. It offers fewer logic resources and a higher ratio of flip-flops to logic resources than a CPLD.

Gate Array, GA

Design implementation with a GA is done with mask design and processing, not by user programming. The manufacturing process has two phases. The first is based on generic masks, or standard masks, and results in an array of uncommitted transistors on the GA IC chip. These chips are later customized as the need arises by defining the metal interconnects between the transistors of the array.

Standard-Cells Based Integrated Circuits, CBIC

In standard-cells based design, all of the commonly used logic cells are designed, characterized, and stored in a standard cell library in a computer. A typical library may consist of a few hundred cells for inverters, NAND gates, NOR gates, latches, flip-flops, etc. Multiple implementation technologies may be used for each cell to provide an adequate choice of electrical characteristics as they may be needed in the design. The characterization of each cell is done for several different categories, including delay-time vs. load capacitance, timing simulation model, circuit simulation model, fault simulation model, cell data for place-and-route, and mask data.

CBIC based design is highly flexible in allowing cells to be placed, modified and connected with a chip area typically 10–15% smaller compared to gate arrays. Each design requires the production of a unique full mask set leading to longer time-to-market and higher non-recurring engineering costs making it a more suitable deign style at high volumes. This design style is sometimes referred to as full-custom design, but in a strict sense it is less than full custom because the cells used are pre-designed and may be utilized in numerous varied chip designs.

Full-Custom Design

In a full-custom design, all the layers of a particular embedded system design are optimized; the optimization includes interconnect lengths and transistor sizes. The design productivity is relatively low, the cost is high, and the time-to-market is long, but excellent performance with optimum size and power consumption can be achieved.

A comparison between the characteristics of various implementation styles is given in Table 1.2.

Table 1.2: Summary of selection criteria for various technology implementation styles.

Criterion	Semicustom			Full Custom
	PLDs	**GAs**	**CBICs**	
Time to market	Short	Medium	Medium	Long
Performance	High	High	High	Very high
Architectural flexibility	Medium to High	High	Higher	Highest
Volume dependence		Low	High	High High
Solution efficiency	Low	High	High	Very high
Application support	Much	Some	Some	None
Design change cost	Medium	High	High	Very high
Development cost	Low	Medium to high	Medium to high	Very high

1.4.3 Design Technologies

A top-down design process comprises progressively refined abstraction levels that include:

- **System Specification**

  At this level, the designer describes the functionality of the system.

- **Behavioral Specifications**

  The designer refines the system specifications by assigning them to processors; this leads to behavioral specifications for each processor.

- **Register Transfer Specifications**

  Behavior specification is further refined by converting the processor behavioral specifications to a set of register transfer (RT), and state machines.

- **Logic Specifications**

  The RT level is further refined into logic specifications expressed through Boolean equations.

  A machine code for general-purpose processors and gate-level netlists for single-purpose processors is then reached.

The design process can be optimized by using appropriate software tools to progress through the levels described. Using libraries of existing implementations could enhance productivity. Such libraries exist at all of the levels, including system-level libraries that might consist of complete systems solving particular problems. Testing correct functionality to prevent time-consuming debugging is most commonly achieved through simulations. At the logic level, gate-level simulators provide output timing waveforms for given input waveforms. General-purpose processor simulators execute machine codes, and hardware description language (HDL) simulators execute RT-level description and provide output timing waveforms for given input waveforms. At the behavioral level, HDL simulators enable hardware/software verification. Model checkers verify the completeness and correctness of the specifications.

1.5 Fuzzy Logic and Embedded Systems

An objective of fuzzy logic has been to make computers *think* like people. Fuzzy logic can deal with the vagueness intrinsic to human thinking and natural language and recognizes that its nature is different from randomness. Using fuzzy logic algorithms could enable machines to *understand* and respond to vague human concepts such as hot, cold, large, small, etc. It also could provide a relatively simple approach to reach definite conclusions from imprecise information.

Almost every application, including embedded control applications, could reap some benefits from fuzzy logic. Its incorporation in embedded systems could lead to enhanced performance, increased simplicity and productivity, reduced cost and time-to-market, along with other benefits. Fuzzy logic has the advantage of modeling complex, nonlinear problems linguistically rather than mathematically and using natural language processing (computing with words). The use of fuzzy logic requires, however, the knowledge of a human expert to create an algorithm that mimics his/her expertise and thinking. Also, studying the stability of a fuzzy system is a demanding task.

Numerous applications, including embedded ones, combine the use of fuzzy logic and neural networks. Neurofuzzy techniques take advantage of both fuzzy logic and neural networks, leading to systems that can:

- Mimic the human decision-making process

- Handle imprecise or vague information

- Learn by example and hence do not require the knowledge of a human expert

- Self-learn and self-organize.

- Process numeric, linguistic, or logical information

Concluding Remarks

Embedded systems have become a ubiquitous part of our daily life at work and play. They have gained importance for productivity and convenience. Industrial control systems, medical instruments, transportation vehicles, bank machines, washing machines, vacuum cleaners, and many other devices now depend on embedded systems. It appears as if our current civilization is built around embedded systems. The demand for higher performance and sophistication of embedded systems is increasing.

Increased market demands require embedded systems to be developed even further at a rapid pace. Fuzzy logic and neural approaches can provide a mechanism for getting the most out of embedded system capabilities and making them more intelligent. They can also accelerate the development cycle and reduce the time-to-market of new products to meet ever-increasing demands. It is important for engineers to know about the capabilities of fuzzy logic and neural networks as possible design approaches from which one may select the most suitable for the problem at hand.

Bibliography

1. I. D. Agranat, "Engineering Web Technologies for Embedded Applications," *IEEE Internet Computing*, 40–45, May/June, 1998.

2. W. Banks and G. Hayward, *Fuzzy Logic in Embedded Microcomputers and Control Systems*, Byte Craft Limited, Waterloo, Ontario, 2001.

3. S. Brown and J. Rose, "FPGA and CPLD Architectures: A Tutorial," *IEEE Design & Test of Computers*, 42–56, Summer 1996.

4. J. P. Calvez, "Performance Assessment of Embedded Hw/Sw Systems," *Proceedings of the International Conference on Computer Design, ICD'95*, 1995, pp. 52–57.

5. M. Chiodo, P. Giusto, A. Jurecska, M. Marelli, H. C. Hsieh, A. S. Vincentelli, and L. Lavagno, "Hardware-Software Codesign of Embedded Systems," *IEEE Micro*, August 1994, pp. 26–36.

6. J. Debardelaben, V. K. Madisetti, and A. J. Gadient, "Incorporating Cost Modeling into Embedded System Design," *IEEE Design & Test of Computers*, July 1997, pp. 24–35.

7. R. E. Filman, "Embedded Internet Systems Come Home," *IEEE Internet Computing*, 52-53, January/February, 2001.

8. D. D. Gajski and F. Vahid, "Specification and Design of Embedded Hardware-Software Systems," *IEEE Design & Test of Computers*, Spring 1995, pp. 53–67.

9. B. H. Lee, "Embedded Internet Systems," *IEEE Internet Computing*, 24-29, May/June, 1998.

10. D. Mlyneck and Y. Leblebici, *Design of VLSI Systems*, vlsi.wpi.edu/webcourse, 1998.

11. D. Mulchandani, "Java for Embedded Systems," *IEEE Internet Computing*, 30–39, May/June, 1998.

12. G. Rozenberg and F. Vandrager, *Lectures on Embedded Systems*, Springer, New York, 1998.

13. M. Schaaf and F. Maurer, "Integrating Java and COBRA," *IEEE Internet Computing*, 72–78, January/February, 2001.

14. M. Schlett, "Trends in Embedded Microprocessor Design," *IEEE Computer*, 44–49, August 1998.

15. D. C. Schmidt, "Middleware Techniques and Optimization for Real-time Embedded Systems," *The 12th International Symposium on System Synthesis*, Boca Raton, Florida, 1-4 November 1999, pp. 12–17.

16. D. Sciuto, "Design Tools for Embedded Systems," *IEEE Design & Test of Computers*, 11–13, April/June 2000.

17. M. J. Sebastian Smith, *Application-Specific Integrated Circuits*, Reading, MA: Addison-Wesley, 1997.

18. Texas Instruments, "Implementation of Fuzzy Logic: Selected Applications," *Technical Report SPRA028*, January 1993.

19. F. Vahid and T. Givargis, *Embedded System Design*, John Wiley & Sons Inc., 2002.

20. A. R. Weiss, "The Standardization of Embedded Benchmarking: Pitfalls and Opportunities," *Proceedings of the IEEE International Conference on Computer Design,* 10-13 October, Austin, Texas, 1999, pp. 492–498.

21. T. Wilmshurst, *An Introduction to the Design of Small-scale Embedded Systems*, Palgrove, New York, 2001.

22. W. Wolf, *Computers as Components: Principles of Embedded Computing System Design*, Morgan Kaufmann Publishers, San Francisco, 2001.

23. W. Wolf, "CAD Techniques for Embedded Systems-on-Silicon," *Proceedings of the IEEE International Conference on Computer Design,* 10–13 October, Austin, Texas, 1999, pp. 291–308.

24. L. A. Zadeh, "Making Computers Think Like People," *IEEE Spectrum,* August 1994, pp. 26–32.

25. L. A. Zadeh, "Fuzzy Logic," *IEEE Computer Magazine*, April 1988, pp. 83–93.

Web Resources

1. EETimes
   www.eet.com/

   This is the Web site of EETimes, a leading weekly newspaper for the Electronics Industry. It provides a section for embedded systems (www.embedded.com/) with news, editorials, articles and downloadable items.

2. Interactive Week
   www.zdnet.com/intweek/

   This the Web site of ZDNet News, a site devoted to the electronics industry. It provides news, technology updates, reviews, downloads and prices. The site is searchable.

3. Embedded Technology
   www.embeddedtechnology.com/

   A site devoted to embedded technology. It provides links to reviews, services and downloads. It enables buying and selling online and gives access to free newsletters and trade publications. The site is searchable.

4. Embedded Systems References
   www.microcontroller.com/

   A Web site intended to provide resources to the professional embedded systems developer. The main categories include: Microcontrollers, Semiconductors, News, Embedded Marketplace, and more. The site is searchable.

5. Programmable Logic Software
   www.optimagic.com/software.html

   This site provides links to sites related to design software for programmable Logic (FPGA, CPLD, and PLD), both free and commercial. They include: Schematic-Based Tools such as:
   VIEWlogic, Mentor Graphics, Cadence Design Systems, OrCAD, ALDEC Active-HDL, Capilano Computing Systems' Design Works, VeriBest, Protel, Tanner Research, Morphologic rapid-development system, MyCAD Logic Synthesis and Optimization including: FPGA Express, FPGA Compiler, Exemplar Logic, Synplicity, VIEWlogic, Cadence Design Systems, Accolade Design Automation, Compass Design Automation Design Verification, Programmable Logic Vendor Software including: Xilinx, Altera, Vantis, Lattice, Lucent, Actel, Cypress Warp2, Atmel FPGA, and University or Research Languages.

6. Compact HTML for Small Appliances
    www.w3.org/TR/1998/NOTE-compactHTML-19980209/

    An article in HTML format about compact HTML for small appliances by
    Tomihisa Kamada, ACCESS Co.,Ltd, Japan. The Compact HTML proposed in
    the document defines a subset of HTML for small information appliances such as
    smart phones, smart communicators, mobile PDAs, etc.

7. Real Time, Embedded and Specialized Systems
    www.omg.org/realtime/

    This is the Web site of the Real-time, Embedded, and Specialized Systems
    Platform Task Force (RTESS PTF). It relates to systems that have one or more of
    the following characteristics: real-time, embedded, fault tolerant, highly avail-
    able, high performance, and safety critical. It provides industry links and
    presentations, and plans to provide white papers on relevant issues.

8. Embedded Microprocessor Benchmark Consortium (EEMBC)
    www.eembc.org

    The Embedded Microprocessor Benchmark Consortium develops and certifies
    benchmarks and benchmark scores to help designers select embedded processors.
    EEMBC was formed in 1997, its membership includes more than 45 leading
    companies.

    According to the statement made on the Web site, every processor submitted for
    bench marking is tested for up to 46 different parameters, each representing a
    different workload and capability in telecom, networking, consumer, office
    automation, and automotive industrial applications. Some sections of the Web
    site require membership.

9. Bluetooth SIG
    www.bluetooth.com/

    A searchable Web site dedicated to Bluetooth technology, which is a worldwide
    specification for low-power, low-cost, short-range wireless networking that
    provides links between mobile computers, mobile phones, other portable
    handheld devices, and connectivity to the internet.

10. Windows NT Embedded Web Site
    www.microsoft.com/windows/embedded/nt

    Microsoft Windows embedded systems Web site. It provides information, news,
    products, etc. It is searchable.

11. Analog Devices
    www.analog.com/

    This is the Web site of Analog Devices. It provides specifications and application notes related to embedded systems and devices, among other information. The site is searchable.

12. Intel
    www.intel.com/research/silicon

    This Web page provides links to Intel's research activities in areas such as nano technology, microprocessors, human computer interface and more. The figures of Chapter 1 were obtained from this Web site.

13. Sun Microsystems
    www.javasoft.com

    This site describes itself as the source for Java Technology. It provides links to information about Java technologies, downloading software tools, solutions, tutorials, and more.

14. Texas Instruments
    http://focus.ti.com/docs/apps/appshomepage.jhtml

    This is the Texas Instruments web page for semiconductor applications including: Audio systems, Broadband Solutions, Digital Control, Video and Imaging, Optical Networking, and Wireless Communications.

15. Free On-line Dictionary of Computing
    www.instantweb.com/~foldoc/

    The On-line Dictionary of Computing, FOLDOC, is a searchable dictionary of acronyms, jargon, programming languages, tools, architecture, operating systems, networking, theory, conventions, standards, mathematics, telecoms, electronics, institutions, companies, projects, products, history; in fact, anything to do with computing. The editor is Denis Howe.

16. Microcontrollers craft a networked future
    www.eetimes.com/story/OEG20010521S0061

    An article by Bernard Cole that appeared in EE Times May 21, 2001. The article indicated that activity is increasing in 8- and 16-bit, and 16- and 32-bit deeply embedded microcontrollers. Even as market attention fixates on 32-bit embedded RISC and DSP in everything from set-top boxes and cell phones to internet-enabled personal digital assistants, the deeply embedded microcontroller space is gaining acreage as opportunities open up. In communications and networking, these parts are playing roles as ancillary processors as they expand the capabilities of existing embedded systems by adding internet ability.

17. Intel MCS 96 Microcontrollers
    www.intel.com/design/mcs96/saback.htm

    Information on the Intel MCS 96 16-bit microcontroller family. It discusses Intel expanding its popular MCS® 96 microcontroller family with the next generation of 16-bit microcontroller architecture.

18. Fujitsu Microcontrollers
    www.fme.fujitsu.com/products/micro/16bit/

    Information from Fujitsu about its spectrum of 16 bit microcontrollers, covering general-purpose and application-specific types with a rich variety of features, including the latest technologies such as on-chip Flash ROM.

19. National Semiconductor Microcontrollers News
    www.national.com/news/0,1737,2000+39,00.html

    This provides news from National Semiconductor about microcontrollers. The site has an archive that is searchable by date and by category.

20. The McClean Report
    www.icinsights.com/prodsrvs/mcclean/mcclean_section10.html

    This is an abstract of Section 10 (Microcontrollers and Digital Signal Processors) of the McClean report, which provided analysis and forecasts for the Integrated Circuit industry.

21. Trends in Embedded Microprocessors Design
    www.realtime-info.be/magazine/98q4/1998q4_p014.pdf

    An article by M. Schlett, Hitachi Europe. It appeared in IEEE Computer, 31, 8, 44–49, 1998.

22. Serial Communications Systems in the Automobile
    www.mcjournal.com/articles/arc100/article4.htm

    An article by Ross Bannatyne, Transportation Systems Group, Motorola Inc. It relates to vehicles, electronic systems architecture, which has become a network of real-time distributed control systems.

23. Aptronix
    www.aptronix.com/tech/

    The Web site of Aptronix, a company specializing in fuzzy logic, expert systems, data mining and neural networks as applied to automation and controls. The page provides links to numerous fuzzy embedded applications.

24. InterScience
    www.intersci.com/projects/fuzzy.html

    According to the the information provided, the compnay has been processing real-time video with the use of Field Programmable Gate Arrays (FPGAs) since 1992. Applications implemented include frame subtraction, convolutions, image-based triggering and compression engines. The page has information about fuzzy logic and links to information on embedded signal processing, automated systems for testing EEPROMs, automated testers, and more.

25. Omron
    http://oeiweb.omron.com/oei/index.htm

    A company specializing in industrial automation and electronic control components. The site is searchable and provides links to information about numerous applications of fuzzy logic in embedded systems applications.

26. Byte Craft Limited Publishing
    www.bytecraft.com/publishing.html

    Two books are available for free in PDF format:
    First Steps with Embedded Systems, and *Fuzzy Logic in Embedded Microcomputers and Control Systems.*

27. Getting Started with Programmable Logic
    http://tutor.al-williams.com/pld-1.htm

    A tutorial on programmable logic from AWS Electronics, League City, Texas, USA. It is presented in HTML format.

28. Links to VLSI Information
    www.mrc.uidaho.edu/vlsi/

    Web links related to VLSI provided by the Microelectronics and Communications Institute, University of Idaho, USA.

29. CPU Info Center
    http://bwrc.eecs.berkeley.edu/CIC/

    A site maintained by Berkeley Wireless Research Center. It provides links to information related to microprocessors, for example: Online Technical Documents, CPU & System, Performance Information, Embedded Microprocessor Information, History of CPUs, and Processor Road-maps.

30. Microelectronics A Tool for Innovation
    www.madess.cnr.it/pubblicazioni/microelectronics_2001.pdf

    A report on the National Italian Project MADESS II. It has extensive reviews and statistics on microelectronics and its applications globally.

31. Conference Proceedings
    http://jamaica.ee.pitt.edu/Archives/ProceedingArchives/

    This site provides links to archives of numerous conferences related to the design of embedded systems, including: Hardware/software Codesign Conference, Design Automation Conference, Design, Automation & Test in Europe Conference, International Conference on Computer Aided Design.

    The proceedings are searchable by author and topic and complete papers are available in PDF format.

CHAPTER **2**

Fuzzy Sets

2.1 Introduction

One can view fuzzy sets as a generalization of classical sets, or crisp sets as they are sometimes called. Classical sets and their operations are particularly useful in expressing classical logic and they lead to Boolean logic and its applications in digital systems. Fuzzy sets and fuzzy operations, on the other hand, are useful in expressing the ideas of fuzzy logic leading to applications such as fuzzy controllers.

The following sections start with a short review of classical sets: their notations and operations. This is followed by fuzzy sets and their operations.

2.2 Classical Sets

A set is defined as a collection of objects that may share certain characteristics. For example, one may define a set of positive integers, a set of students with passing grades, and a set of honest politicians. Each individual object is referred to as an *element* or *member* of the set. In a classical set an object x is either a member of a given set A (expressed as $x \in A$) or not a member (expressed as $x \notin A$); partial membership is not allowed.

There are numerous ways to define a set:

- One may specify the properties of its elements. For example,

    $A = \{x | x \text{ is an odd number} <10\}$

- One may list all the members of the set. For example,

    $A = \{1, 3, 5, 7, 9\}$

- One may use a formula to define the set. For example,

    $A = \{x_i = x_i + 1, i = 1,...,5, \text{ where } x_i = 1\}$

- The set could also be defined as the result of a logical operation. For example,

    $A = \{x | x \text{ is an element that belongs to } B \text{ OR } C\}$

- A membership function, μ, can be used to define a set.

$\mu_A(x) = 1$ if $x \in A$, and

$\mu_A(x) = 0$ if $x \notin A$ for all values of x.

Example

Let all the numbers under consideration, i.e. the universe of discourse, be defined as

$\{1, 2, 3, 4, 5, 6, 7, 8, 9, 10\}$.

Then, the set of odd numbers can be expressed as

$\{(1,1), (2,0), (3,1), (4,0), (5,1), (6,0), (7,1), (8,0), (9,1), (10,0)\}$.

Where each member of the universe of discourse is associated with a membership value in the form (#, μ). The numbers 1, 3, 5, 7, and 9 are associated with $\mu = 1$ because they form the set of odd numbers extracted from the universe of discourse.

This method of defining a set can be easily extended to define a fuzzy set by allowing partial membership.

Universal Set

The set that consists of all the elements of interest for a particular application (the universe of discourse) is referred to as the *universal set*. It is the mother of all sets; any set that is not a universal set is a subset. One may write $A \subset I$ to mean that a set A (any set) is actually a subset of the universal set I.

Empty Set

A set that has no elements is referred to as an *empty set* and is denoted by the symbol $\varnothing$. It is a shell that represents a description that does not happen to fit anything. (Contrary to what some may think, the set of "Honest Politicians" is not an example of an empty set!) An empty set is a subset of any other set, i.e. $\varnothing \subseteq A \subseteq I$.

Power Set

The set which consists of all possible subsets of a given set A is called a *power set* and is denoted as $P(A) = \{x | x \subseteq A\}$

Example

If $A = \{1, 2, 3\}$, then

$P(A) = \{\varnothing, \{1\}, \{2\}, \{3\}, \{1,2\}, \{1,3\}, \{2,3\}, \{1, 2, 3\}\}$

Cardinality

The number of elements of a given set is called the cardinality of the set. The cardinality of a set A is denoted by $\# A$ or $|A|$. The cardinality is a finite number for finite sets.

Example

If $A = \{1, 2, 3, 5, 7\}$, then $\# A = 5$.

2.3 Set Operations

Sets can be manipulated through numerous operations such as the complement, union, intersection, subtraction, and cartesian product. These will be defined and explained in the following sections.

COMPLEMENT

The COMPLEMENT, or ABSOLUTE COMPLEMENT of a given set A is denoted by $\overline{A}$. It is defined by

$$A = \{ \, x \, | \, x \in I \ \text{and} \ x \notin A \, \}.$$

It is demonstrated graphically in Figure 2.1.

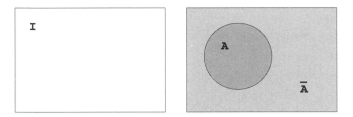

Figure 2.1: The complement of set A.

If the membership function of set A is $\mu_A(x)$ and that of $\overline{A}$ is $\mu_{\overline{A}}(x)$, then one can write

$$\mu_{\overline{A}}(x) = 1 - \mu_A(x).$$

This is illustrated graphically in Figure 2.2.

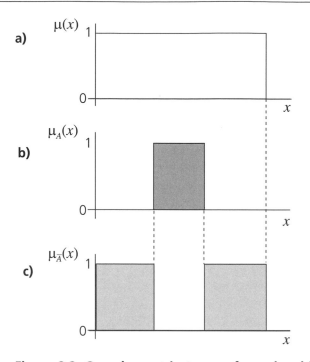

Figure 2.2: Complement in terms of membership function.
a) the universe of discourse
b) set *A*
c) the complement of *A*

Example

Let $I = \{1, 2, 3, 4, 5, 6, 7, 8, 9, 10\}$

If $A = \{(1,1), (2,1), (3,1), (4,1), (5,1), (6,0), (7,0), (8,0), (9,0), 10,0)\}$

$\qquad = \{1, 2, 3, 4, 5\}$

Then, $\bar{A} = \{(1,0), (2,0), (3,0), (4,0), (5,0), (6,1), (7,1), (8,1), (9,1), (10,1)\}$

$\qquad = \{6,7,8,9,10\}.$

Example

If $I = \{0, 1\}$ and $A = \{1\}$, then

$\bar{A} = \{0\}$

The complement of the complement of *A* is *A* itself. This property is known as *involution,* which is similar to the logical double negation.

UNION

The UNION of sets A and B is defined by

$$A \cup B = \{ \, x \, | \, x \in A \text{ or } x \in B \, \}.$$

It is illustrated graphically in Figure 2.3.

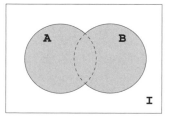

Figure 2.3: The union of two sets.

If the membership function of set A is $\mu_A(x)$ and that of set B is $\mu_B(x)$, then the membership of the set resulting from $A \cup B$ can be represented graphically as shown in Figure 2.4.

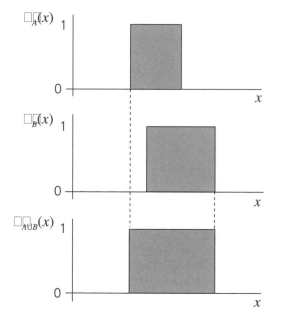

Figure 2.4: Union in terms of membership function.

Example

Let $A = \{1, 2, 3, 4, 5, 6\}$ and $B = \{3, 5, 7, 8\}$, then

$A \cup B = \{1, 2, 3, 4, 5, 6, 7, 8\}$.

Example

If $A = \{1\}$, $B = \{0\}$, and $I = \{0,1\}$, then

$A \cup B = I$

INTERSECTION

The INTERSECTION of sets A and B is defined by

$$A \cap B = \{ x \,|\, x \in A \text{ and } x \in B \}.$$

It is illustrated graphically in Figure 2.5, and in terms of the membership function Figure 2.6.

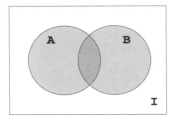

Figure 2.5: Intersection of two sets.

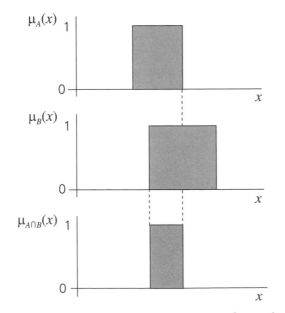

Figure 2.6: Intersection in terms of membership function.

Example

Let $A = \{1, 2, 3, 4, 5, 6\}$ and $B = \{2, 4, 6, 8, 10\}$, then

$A \cap B = \{2, 4, 6\}$.

Example

Let $A = \{1\}$ and $B = \{0\}$, then

$A \cap B = \varnothing$.

SUBTRACTION

The difference of sets A and B is a set that consists of all the elements which belong to A, but do not belong to B. It can be expressed as

$$A - B = A - A \cap B$$
$$= \{\, x \mid x \in A \text{ and } x \notin B \,\}.$$

The operation is illustrated in Figure 2.7.

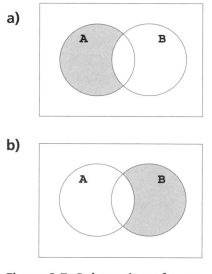

Figure 2.7: Subtraction of two sets.
a) $A - B$
b) $B - A$

Cartesian Product

The Cartesian product of sets A and B is defined as $A \times B \{(a,b)| \; a \in A, b \in B\}$.

Example

Let $A = \{1,2\}$ and $B = \{3, 4, 5\}$, then

$A \times B = \{(1,3), (1,4), (1,5), (2,3), (2,4), (2,5)\}$, and

$B \times A = \{(3,1), (3,2), (4,1), (4,2), (5,1), (5,2)\}$.

Table 2.1 summarizes some basic properties of classical set operations. It useful to observe the duality some properties exhibit by replacing $\varnothing, \cup, \cap$ with $I, \cap, \cup$, respectively. It is also useful to remember that the law of contradiction and the law of excluded middle are fundamental to classical logic (and our way of thinking) to contrast with fuzzy logic later.

Table 2.1: Some basic properties of classical set operations.

(note that the negation operation can be expresses as $\overline{A}$ or A')

Law of contradiction	$A \cap \overline{A} = \varnothing$
Law of excluded middle	$A \cup \overline{A} = I$
De Morgan's laws	$(A \cap B)' = A' \cup B'$
	$(A \cup B)' = A' \cap B'$
Involution (Double negation)	$\overline{\overline{A}} = A$
Commutative	$A \cap B = B \cap A$
	$A \cup B = B \cup A$
Associative	$A \cap (A \cap B) = (A \cap B) \cap C$
	$A \cup (B \cup C) = (A \cup B) \cup C$
Distributive	$A \cap (B \cup C) = (A \cap B) \cup (A \cap C)$
	$A \cup (B \cap C) = (A \cup B) \cap (A \cup C)$

2.4 Boolean Logic

The basic building blocks of digital systems, including computers and embedded systems, are electronic switching devices, such as transistors, or algorithms. Since such devices assume two conditions: ON or OFF, the universe of discourse can be represented by {0, 1}. A set with one element {0}, referred to as a *singleton*, can represent the OFF condition, another singleton{1}can represent the ON condition. Operations on such sets are special cases of the general set operations, they are referred to as binary or Boolean operations. The correspondence between basic set operations and binary logic operations are outlined in Table 2.2.

Table 2.2: Basic binary operations.

Operation	Symbol	Corresponding Set Operation
NEGATION	$\neg$	COMPLEMENT
OR	$+, \vee$	INTERSECTION
AND	$\cdot, \wedge$	UNION

The circuits that perform the AND, OR, and NEGATION operations are referred to as logic gates. They are available as IC chips. Numerous circuits built using such gates are also available as IC chips, the number of gates per chip defines the level of integration as discussed in Chapter 1.

2.5 Basic Concepts of Fuzzy Sets

A fuzzy set is a set where degrees of membership between 1 and 0 are allowed; it allows partial membership. Fuzzy sets can thus better reflect the way intelligent people think. For example, an intelligent person will not classify people as either friends or enemies; there is a range between these two extremes. Not recognizing that there are degrees in every trait can lead to erroneous decisions.

Vague human expressions such as *tall, hot, cold*, etc. can be expressed by fuzzy sets of the form

$$A = \{ (x, \mu_A(x)) | x \in X \}$$

where X represents the universe of discourse and $\mu_A(x)$ assumes values in the range from 1 to 0.

31

Example

Let the values of temperature in °C under consideration be

$T = \{0, 5, 10, 15, 20, 25, 30, 35, 40\}$.

Then, the term *hot* can be defined by a fuzzy set as follows

HOT = $\{(0,0), (5,0.1), (10,0.3), (15,0.5), (20,0.6), (25,0.7), (30,0.8), (35,0.9), (40,1.0)\}$.

This fuzzy set reflects the point of view that 0 °C is not hot at all, 5, 10, and 15 °C are somewhat hot, and 40 °C is indeed hot. Another person could have defined the set differently.

2.6 Other Representations of Fuzzy Sets

Fuzzy sets can be described in numerous ways; all of them allow partial membership to be expressed. The ordered pair method introduced in the previous section appears in a different format as follows

$$A = \mu_1/x_1 + \mu_2/x_2 + \mu_3/x_3 + \dots$$

The symbol / here does not denote division, nor does the symbol + denote summation. The summation symbol is used to connect the terms and thus it means a union of single-term subsets.

Example

Let $X = \{ x_1, x_2, x_3, x_4 \}$. One can define a fuzzy set as:

$$A = 0.8/x_1 + 0.4/x_2 + 0.1/x_3 + 0.9/x_4 \ .$$

Fuzzy sets can also be defined by assigning a continuous function to describe the membership either analytically or graphically. Some commonly used membership functions are illustrated in Figure 2.8. The triangular membership function in Figure 2.8-a can also be expressed as

$$\mu(x) = a\ (b\text{-}x)/(b\text{-}c) ; \qquad b \geq x \leq c$$

$$= a\ (d\text{-}x)/(d\text{-}c) ; \qquad c \geq x \leq d$$

$$= 0 ; \qquad \text{otherwise}$$

The trapezoidal function shown in Figure 2.8-b can be expressed as:

$$\mu(x) = a\ (b\text{-}x)/(b\text{-}c) ; \qquad b \geq x \leq c$$

$$= a ; \qquad c \geq x \leq d$$

$$= a\ (e\text{-}x)/(e\text{-}d) ; \qquad d \geq x \leq e$$

$$= 0 ; \qquad \text{otherwise}$$

The Gaussian function of Figure 2.8-c can be expressed as:

$$\mu(x) = a \exp(- x-b)^2/2\sigma^2)$$

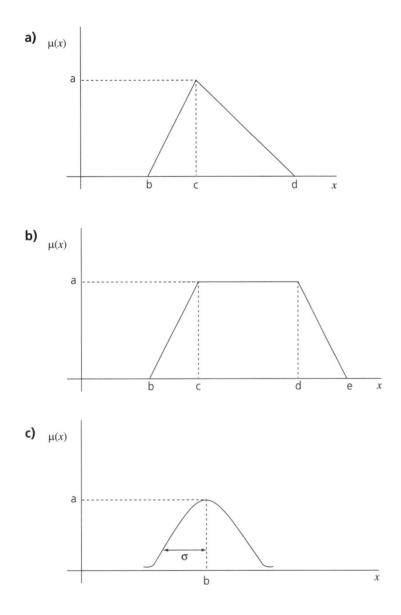

Figure 2.8: Common membership functions.
a) Triangular
b) Trapezoidal
c) Gaussian

Example

Figure 2.9 shows three sets defined graphically to represent the fuzzy sets SLOW, MEDIUM, and FAST to reflect a way of thinking about values of speed in the range of 0 to 40 km/hr.

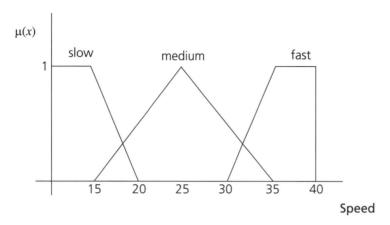

Figure 2.9: Possible graphical representation of fuzzy sets.

Example

The vague expression *around ten* can be expressed as a fuzzy set graphically as shown in Figure 2.10.

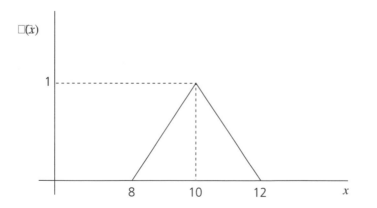

Figure 2.10: *Around ten* expressed as a fuzzy set graphically.

2.7 Determination of Membership Functions

Discrete and continuous membership functions of a fuzzy set are intended to capture a person's thinking. The fuzzy set in the example of Section 2.5 was defined based on one's thinking of what coldness is. Fuzzy membership functions can still be determined subjectively in practical problems based on an expert's opinion. In such a situation one can think of membership functions as a technique to formalize empirical problem solving that is based on experience rather than the knowledge of theory. The expert's way of thinking can be captured either directly or through a special algorithm. Such determination could become more focused by physical measurements if the need arises. Available frequency histograms and other probability data can also help in constructing the membership function. It is important, however, to note that membership function values, or grades of membership, are not probabilities and they do not have to add to 1. Membership construction can be further simplified by selecting their form from the smaller family of the commonly used ones, such as those shown in Figure 2.8.

2.8 Fuzzy Sets Properties

Empty fuzzy set

A fuzzy set is referred to as empty if and only if the value of the membership function is zero for all possible members under consideration. In a short hand form this statement would read

$$A = \emptyset \text{ iff } \mu_A(x) = 0 \; \forall x \in X \, . \qquad \text{(iff and } \forall \text{ are short hand forms for } \textit{if and only if} \text{ and } \textit{for all values of}, \text{ respectively)}.$$

Universal fuzzy Set

A fuzzy set is universal if and only if the value of the membership function is one for all members under consideration.

Equal fuzzy sets

Two fuzzy sets A and B are said to be equal iff $\mu_A(x) = \mu_B(x)$ for all $x \in X$.

α -cuts

A fuzzy set may be completely characterized by its α-cuts, defined as follows

strong α-cuts: $\qquad A_\alpha = \{ \, x \, | \, \mu_A(x) > \alpha \, \}; \; \alpha \in [0,1]$

weak α-cuts: $\qquad A_{\bar{\alpha}} = \{ \, x \, | \, \mu_A(x) \geq \alpha \, \}; \; \alpha \in [0,1]$

Thus, an α-cut is a crisp set that consists of all the elements of a fuzzy set whose membership functions have values greater than a specified value α, or greater than or equal to a specified value; the first condition leads to strong α-cuts and the second to weak α-cuts. All the cuts of a fuzzy set form a family of crisp sets.

Example

Let $A = 0.2/1 + 0.5/2 + 0.6/3 + 1/4 + 0.7/5 + 0.3/6 + 0.1/7$.

Then, the weak α-cuts for $\alpha \in [0,1]$ from 0.1 to 1 with step width of 0.1 are as follows

$A_{0.\bar{1}} = \{1, 2, 3, 4, 5, 6, 7\}$

$A_{0.\bar{2}} = A_{0.\bar{3}} = \{1, 2, 3, 4, 5, 6\}$

$A_{0.\bar{4}} = A_{0.\bar{5}} = \{2, 3, 4, 5\}$

$A_{0.\bar{6}} = \{3, 4, 5\}$

$A_{0.\bar{7}} = \{4, 5\}$

$A_{0.\bar{8}} = A_{0.\bar{9}} = A_{1.\bar{0}} = \{4\}$

Support

The support of a fuzzy set A is a crisp set supp(A) of all $x \in X$ such that $\mu_A(x) > 0$. It is a strong α-cut for $\alpha = 0$. The element $x \in X$ at which $\mu_A(x) = 0.5$ is referred to as the *cross-over point*. A fuzzy set whose support is a single element in X with $\mu_A(x) = 1$ is referred to as a *fuzzy singleton*.

Core

The core of a fuzzy set A is a crisp set core(A) of all $x \in X$ such that $\mu_A(x) = 1$. The core of a fuzzy set may be an empty set.

Height

The height, h(A) of a fuzzy set A is the largest value of μ_A for which the α-cut is not empty. In other words, it is the largest value of the membership function attained by an element in the set. A fuzzy set with h(A) = 1 is referred to as *normal*, otherwise it will be referred to as *sub-normal*.

The concepts of α-cuts, support, core, and height are illustrated in Figure 2.11.

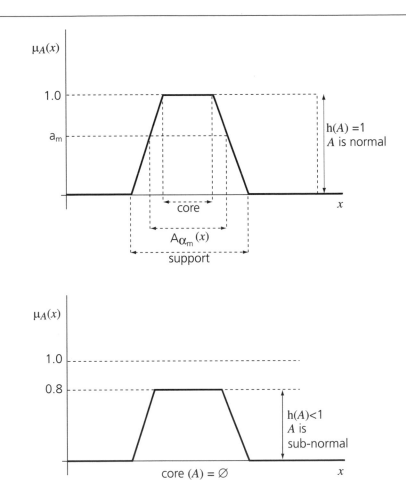

Figure 2.11: A graphical illustration of α-cuts, support, core, and height.

2.9 Operations on Fuzzy Sets

2.9.1 Logic Operations

The three basic logic operations on classical (crisp) sets were introduced in Section 2.2. The definitions of sets were generalized in Section 2.5 leading to fuzzy sets upon which similar operations can be performed. The generalizations of operations on sets to operations on fuzzy sets are not unique. The ones introduced here are referred to as the *standard fuzzy set operations*. They are the operations most commonly used in engineering applications.

COMPLEMENT

The absolute complement of a fuzzy set A is denoted by $\overline{A}$ and its membership function is defined by

$$\mu_{\overline{A}}(x) = 1 - \mu_A(x) \text{ for all } x \in X$$

Example

Let $A = 0.7/1 + 1/1 + 0.6/3 + 0.2/4 + 0/5$.

Then, $\overline{A} = 0.3/1 + 0/2 + 0.4/3 + 0.8/4 + 1/5$.

UNION

The union of two fuzzy sets A and B is a fuzzy set whose membership function is defined by

$$\mu_{A \cup B}(x) = \max\ [\mu_A(x), \mu_B(x)\ ]$$

Example

Let $A = 0.3/1 + 0/2 + 0.4/3 + 0.8/4 + 1/5$, and

$B = 0.2/1 + 0.3/2 + 0.1/3 + 0.2/4 + 0.4/5$.

Then,

$$A \cup B = 0.3/1 + 0.3/2 + 0.4/3 + 0.8/4 + 1/5$$

INTERSECTION

The intersection of two fuzzy sets A and B is a fuzzy set whose membership function is defined by

$$\mu_{A \cap B}(x) = \min[\mu_A(x), \mu_B(x)].$$

Example

Let $A = 0.3/1 + 0/2 + 0.4/3 + 0.8/4 + 1/5$, and

$B = 0.2/1 + 0.3/2 + 0.1/3 + 0.2/4 + 0.4/5$

Then,

$$A \cap B = 0.2/1 + 0/2 + 0.1/3 + 0.2/4 + 0.4/5$$

Some properties of fuzzy set operations are given in Table 2.3. It is particularly important to observe that the laws of contradiction and excluded middle are not applicable. How far a set deviates from such laws can be a measure of fuzziness. A graphical illustration of basic fuzzy sets operations is presented in Figure 2.12.

Table 2.3 Some properties of fuzzy sets operations

Law of contradiction	$A \cap \bar{A} \neq \varnothing$
Law of excluded middle	$A \cup \bar{A} \neq I$
De Morgan's laws	$\left(A \cap B\right)' = A' \cup B'$ $\left(A \cup B\right)' = A' \cap B'$
Involution (Double negation)	$\bar{\bar{A}} = A$
Commutative	$A \cap B = B \cap A$ $A \cup B = B \cup A$
Associative	$A \cap \left(A \cap B\right) = \left(A \cap B\right) \cap C$ $A \cup \left(B \cup C\right) = \left(A \cup B\right) \cup C$
Distributive	$A \cap \left(B \cup C\right) = \left(A \cap B\right) \cup \left(A \cap C\right)$ $A \cup \left(B \cap C\right) = \left(A \cup B\right) \cap \left(A \cup C\right)$

a)

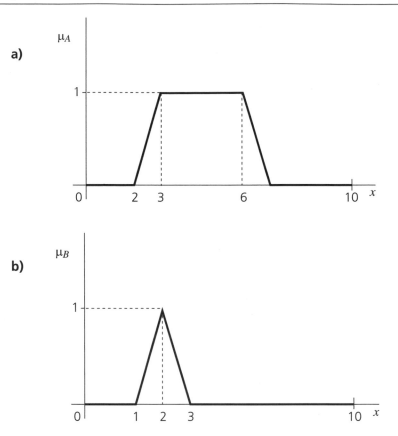

b)

Figure 2.12: Graphical illustration of basic fuzzy sets operations
a) Set *A*   b) Set *B*

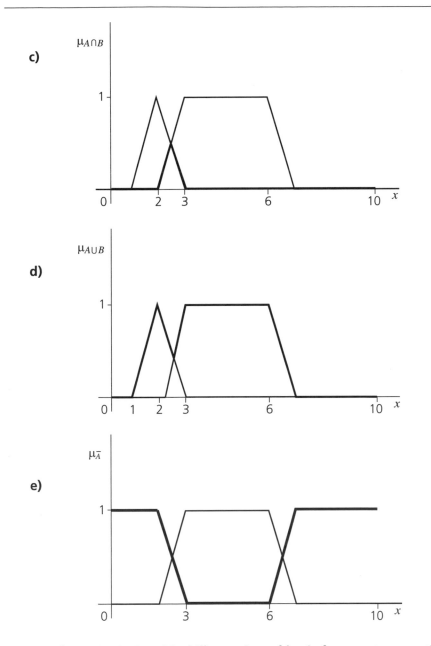

c)

d)

e)

Figure 2.12: Graphical illustration of basic fuzzy sets operations (cont.)

    c) $A \cap B$ **d)** $A \cup B$

    e) $\overline{A}$

2.9.2 Algebraic Operations on Fuzzy Sets

Cartesian multiplication

The Cartesian multiplication of two sets A and B is a fuzzy set C such that

$$C = A \times B$$
$$= \left\{ \mu_C(x)/(a,b) \middle| a \in A, b \in B, \mu_C(c) = \min\left[\mu_A(a), \mu_B(b)\right] \right\}$$

Example

Let A = 0.2/3 + 1/5 + 0.5/7, and
$\quad$ B = 0.8/2 + 0.3/6.

Then,

$$A \times B = \min[0.2,0.8]/(3,2) + \min[0.2,0.3]/(3,6)$$
$$+ \min[1,0.8]/(5,2) + \min[1,0.3]/(5,6)$$
$$+ \min[0.5,0.8]/(7,2) + \min[0.5,0.3]/(7,6)$$
$$= 0.2/(3,2) + 0.2/(3,6) + 0.8/(5,2) + 0.3/(5,6)$$
$$+ 0.5/(7,2) + 0.3/(7,6).$$

Algebraic multiplication

The algebraic product of two fuzzy sets A and B leads to a fuzzy set C such that

$$AB = \left\{ \mu_A(a)\mu_B(b)/x \middle| x \in A, x \in B \right\}.$$

Example

Let A = 0.2/3 + 1/5 + 0.5/7, and

$\quad$ B = 0.1/3 + 0.3/7 + 0.2/8.

Then, AB = (0.2)(0.1)/3 + (1)(0)/5 + (0.5)(0.3)/7 + (0)(0.2)/8

$$= 0.02/3 + 0.15/7$$

Exponent

Raising a set A to the power of α is a special case of algebraic multiplication. It is defined by

$$A^\alpha = \left\{ (\mu_A(x))^\alpha /x \middle| x \in A \right\}.$$

Example

Let $A = 0.5/4 + 1/5 + 0.4/6$.

Then, $A^2 = (0.25)^2/4 + (1)^2/5 + (0.4)^2/6$

$$= 0.25/4 + 1/5 + 0.16/6, \text{ and}$$

$$A^{0.5} = (0.5)^{0.5}/4 + (1)^{0.5}/5 + (0.4)^{0.5}/6$$

$$= 0.7/4 + 1/5 + 0.63/6.$$

As pointed out earlier, a fuzzy set A can be expressed as a linguistic concept such as *hot, cold, young, old*, etc. The result of using nested linguistic modifiers such as *very, very very*, etc. can be expressed using A^α.

Example

Let the range of temperature under consideration be 5, 10, 20, 30, and 40°C. Let also the set A represent the concept "hot" from a given point of view:

$A = 0.1/5 + 0.2/10 + 0.4/20 + 0.6/30 + 0.7/40$.

Then, *very hot* could be expressed by

$A^2 = 0.01/5 + 0.04/10 + 0.16/20 + 0.36/30 + 0.49/40$.

It follows that *very very hot* can be expressed by $(A^2)^2 = A^4$.

Concentration and Dilation

These two operations are unique to fuzzy sets; they do not have counterparts in classical sets. The concentration operation is defined as:

$$\text{CON}(A) = A^2$$

The dilation process is defined as:

$$\text{DIL}(A) = A^{0.5}$$

Example

Let $A = 0.5/4 + 1/5 + 0.4/6$.

Then, $\text{CON}(A) = 0.25/4 + 1/5 + 0.16/6$, and

$$\text{DIL}(A) = 0.7/4 + 1/5 + 0.63/6.$$

The operations concentration and dilation could be composed with themselves, eg, $\text{CON}^2(A) = A^4$,

and $\text{DIL}^2(A) = A^{\frac{1}{4}}$. In general one can write

$$\text{CON}^\alpha(A) = A^{2\alpha}, \text{ and}$$

$$\text{DIL}^\alpha(A) = A^{0.5\alpha}.$$

with α being an integer ≥ 2.

The CON operation reduces the value of $\mu_A(x)$ except for $\mu_A(x) = 1$. As $\alpha \to \infty$, the membership function becomes two-valued: 1 and 0. On the other hand, the DIL operation increases the value of $\mu_A(x)$ except for $\mu_A(x) = 1$. As $\alpha \to \infty$, $\mu_A(x) \to 1$.

The effect of the CON operation on the membership function is illustrated graphically in Figure 2.13.

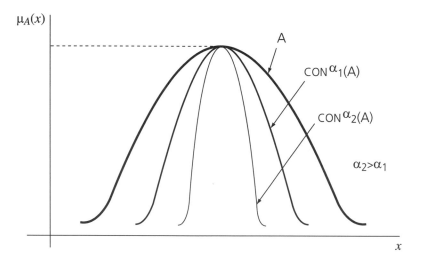

Figure 2.13: Graphical illustration of the effect of the CON operation on the membership function.

Algebraic sum

The algebraic sum of two sets A and B is a fuzzy set with a membership function given by:

$$\mu_A(x) + \mu_B(x) - \mu_A(x)\mu_B(x)$$

where the symbol + here denotes algebraic summation.

Example

Let C be the algebraic summation of two fuzzy sets A and B expressed by:

$C = A + B$,

$A = 0.2/1 + 1/2 + 0.4/3$, and

$B = 0.3/1 + 0.4/2 + 0.2/7$.

Then, the set C would consist of four members: 1, 2, 3, and 7 with membership functions $\mu_C(x_1), \mu_C(x_2), \mu_C(x_3), \mu_C(x_4)$ respectively. They can be determined as follows:

$$\mu_C(x_1) = 0.2 + 0.3 - (0.2)(0.3)$$
$$= 0.44$$

$$\mu_C(x_2) = 1 + 0.4 - (1)(0.4)$$
$$= 1$$

$$\mu_C(x_3) = 0.4 + 0 + (0)(0.4)$$
$$= 0.4$$

$$\mu_C(x_4) = 0.2 + 0 + (0)(0.2)$$
$$= 0.2$$

Accordingly, $C = 0.44/1 + 1/2 + 0.4/3 + 0.2/7$.

(One should remember that the symbol + here is not algebraic summation.)

Bounded sum

The symbol $\oplus$ denotes the bounded sum of two fuzzy sets. The operation leads to a fuzzy set with a membership function defined by:

$$\mu_{A \oplus B}(x) = \min\left[1, \left(\mu_A(x) + \mu_B(x)\right)\right]$$

Example

Let $A = 0.5/3 + 1/4 + 0.8/5$, and

$\quad B = 0.2/3 + 0.4/5$.

Then, $A \oplus B = 0.7/3 + 1/4 + 1/5$.

Bounded difference

The symbol $\ominus$ denotes the bounded difference of two fuzzy sets. The operation leads to a fuzzy set with a membership defined by:

$$\mu_{A \ominus B}(x) = \min\left[1, \left(\mu_A(x) + \mu_B(x)\right)\right]$$

Example

Let $A = 0.5/3 + 0.3/5 + 0.7/8$, and

$\quad B = 0.4/3 + 0.1/5$.

Then, $A \ominus B = \mu(x_1)/3 + \mu(x_2)/5 + \mu(x_3)/8$

where $\quad \mu(x_1) = \min[1,(0.5 - 0.4)] = 0.1$,

$\qquad \mu(x_2) = \min[1,(0.3 - 0.1)] = 0.2$, and

$\qquad \mu(x_3) = \min[1,(0.7 - 0.0)] = 0.7$.

hence, $\quad A \ominus B = 0.1/3 + 0.2/5 + 0.6/8$.

Some identities that relate sets and algebraic operations are given in Table 2.4. They will be of particular importance when discussing some of the embedded applications of fuzzy sets.

Table 2.4: Identities relating sets and algebraic operations.

$$\mu_A(x) \wedge \mu_B(x) = \mu_B(x) \ominus \left(\mu_B(x) \ominus \mu_A(x)\right)$$

$$\mu_A(x) \vee \mu_B(x) = \mu_A(x) + \left(\mu_B(x) \ominus \mu_A(x)\right)$$

$$\left|\mu_A(x) - \mu_B(x)\right| = \left(\mu_A(x) \ominus \mu_B(x)\right) + \left(\mu_B(x) \ominus \mu_A(x)\right)$$

Fuzzification

Fuzzification is the operation of transforming a crisp set to a fuzzy set, or a fuzzy set to a fuzzier set. The operation translates crisp input or measured values into linguistic concepts. This, in a way, is similar to what people may do in numerous situations to reach a decision. For example, if one is told that the temperature is going to be 10 °C, one translates this crisp input value into a linguistic concept such as *mild, cold*, or *warm* according to one's inclination, then reaches a decision about the need to wear a jacket or not. If one fails to fuzzify (for example, due to lack of familiarity with the Celsius temperature scale) then the decision process cannot continue or a possibly erroneous decision would be reached. So, you have been fuzzifying all along (without knowing it) whenever you made correct decisions.

A common fuzzification algorithm of a set $A = \left\{ \mu_i \big/ x_i \, | x_i \in X \right\}$ is performed by keeping μ_i constant and transforming x_i to a fuzzy set that depicts the expression *about* x_i, $K(x_i)$. The fuzzy set $K(x_i)$ is referred to as the kernel of fuzzification. Fuzzified set A is expressed as

$$\sim A = \mu_1 K(x_1) + \mu_2 K(x_2) + \dots + \mu_n K(x_n)$$

where the symbol $\sim$ means *fuzzified*.

The method is referred to sometimes as *support fuzzification* (s-fuzzification) in order to distinguish it from another possible fuzzification method referred to as *grade fuzzification* (g-fuzzification) in which x_i is kept constant and μ_i is expressed as a fuzzy set.

It is interesting to observe that both the dilation and fuzzification operations lead to a fuzzier set. However, dilation of a crisp set yields the same crisp set, which is not typically the case with the fuzzification operation.

It is useful to devise a measurement to determine how fuzzy a set is. All sets that allow partial membership are fuzzy, but some are fuzzier or can be made fuzzier than others. One can intuitively suggest that the sharper the boundaries, i.e. the closer the values of the membership to 1 or 0, the less fuzzy a set is. A more formal measure of fuzziness could be desirable to define. A possible measure of fuzziness can be put forward by evaluating the deviation of a set from its complement, i.e. $\left|\mu_A(x)-\mu_{\bar{A}}(x)\right|$.

Such a difference can be expressed as

$$\left|\mu_A(x)-\left(1-\mu_A(x)\right)\right|=\left|2\mu_A(x)-1\right|$$

The largest difference occurs when $\mu(x)$ is at its largest value, which is 1, implying that the largest difference would be 1. The degree of the lack of distinction between a fuzzy set and its complement can then be expressed by $1-\left|2\mu_A(x)-1\right|$.

A fuzzy measure $f(A)$ for a finite set can then be defined by

$$f(A)=\sum_{x\in X}\left(1-\left|2\mu_A(x)-1\right|\right)$$

Note that $f(A) \geq 0$ for all fuzzy sets and $f(A) = 0$ for all crisp sets. Also $f(A)$ is at its maximum when $\mu(x) = 0.5$ since this is where the membership function is farthest away from both 1 and 0.

Concluding Remarks

Fuzzy sets are the tools which convert the concepts of fuzzy logic into algorithms leading to applications. They are used to express *precisely* what one means by vague expressions such as *hot, cold, tall*, and *short*. By allowing partial membership, fuzzy sets can provide computers with algorithms that extend their binary logic and enable them to make human-like decisions. The term fuzzy in this context does not mean imprecise, but exactly the opposite.

It may appear difficult to reconcile the claim that the objective of fuzzy sets is to enable computing with words with the fact that fuzzy sets are of a mathematical nature. Do humans think in terms of triangular membership functions, Cartesian multiplications, etc.? Certainly not, but one may think of fuzzy sets and the associated mathematics as the media through which the way we think is transferred to a computer, rather than trying to accommodate our thinking to the computer needs. Of course, it would have been more efficient to have a means for direct transfer of thoughts without the intermediate stage of mathematics. The avoidance of mathematics occurs in the way to describe a system.

Attempting to transfer our way of thinking into fuzzy set formulation could have an interesting side effect. It gives us an opportunity to ponder about our own thoughts and actions and reflect on the wisdom of our choices and judgements.

Bibliography

1. G. J. Klir and T. A. Floger, *Fuzzy sets, Uncertainty, and Information*, Prentice Hall International, London, UK, 1988.

2. G. J. Klir, U. H. St. Clair, and B. Yuan, *Fuzzy Set Theory*, Prentice Hall, Upper Saddle River, NJ, 1997.

3. S. Lipschuhutz, *Discrete Mathematics*, McGraw-Hill, Toronto, 1976.

4. K. H. Rosen, *Discrete Mathematics and its Applications*, McGraw-Hill, Toronto, 1999.

5. Y. Shirai, T. Yamakawa, and F. Ueno, "A CAD-oriented Synthesis of Fuzzy Logic Circuits," *Systems Computers Controls*, 15, 76–83, 1984.

6. L. A. Zadeh, "Fuzzy Sets," *Information Control* 8, 338–353, 1965.

7. L. A. Zadeh, "A Fuzzy-Set Theoretic Interpretation of Linguistic Hedges," *J. Cybern*, 2, 3, 4–34, 1972.

8. L. A. Zadeh, "Outline of a New Approach to the Analysis of Complex Systems and Decision Processes," *IEEE Trans. Sys. Man Cybern*, SMC-3, 28–44, 1973.

Web Resources

1. Notes on Set Theory
    www.math.csusb.edu/notes/sets/sets.html

    A set of notes presented in HTML format by Nikos Drakos, Computer Based Learning Unit, University of Leeds. It discusses basic topics related to Set Theory including: notations and terminology, set operations, and Venn diagrams.

2. Set Theory
    www.math.niu.edu/~rusin/known-math/index/03EXX.html

    An article from the Mathematical Atlas which evolved from a much more informal collection of mathematical pointers collected by Prof. Dave Rusin of Northern Illinois University. It provides extensive Web links to topics such as the history of set theory, applications and related fields, an annotated list of textbooks, and a list of software and tables. It also provides links to Web sites with similar interests.

3. Sets, Logic and Foundations and Philosophy of Mathematics
    www.mathlinks.info/em035_set_log_foun_philo.htm

    A collection of Web links from Math.Links maintained by Ronald N. Gibson. The links are classified into categories that include: Academic Programs in Logic and Mathematical Foundations, Articles, Courses, Lectures, Texts, Tutorials, Journals, Organizations, and more.

4. Set Theory Page
    www.cis.syr.edu/~sanchis/setory.html

    A collection of links to Web sites by Prof. Luis Sanchis, Syracuse University.

    Web sites listed include:

    Situated Set Theory, Set Theory, Bounded Set Theory, Holmes: New Foundations Home Page, Constructive Set Theory, Programming with Sets, Forster Bibliography, E-mail Addresses of Set Theorists, Homepages of Set Theorists, Beginnings of Set Theory, Logic Eprints, Fixed Point Theory, and much more.

5. Java Set Theory Machine
    www.jaytomlin.com/music/settheory/

    A page maintained by Jay Tomlin. It provides an interactive Java applet related to the Musical Set Theory. It also provides a short tutorial about the topic. The source code is available for downloading.

6. Logic and Set Theory
   www.humboldt.edu/~mef2/logicsites.html

   An extensive collection of web links maintained by Professor Martin Flashman, Department of Mathematics, Humboldt State University, Arcata, California, USA. The links are organized in categories including: Logic, Set Theory, Turing Machine, History, and Major References.

7. The Mathematics of Set Theory
   www.jboden.demon.co.uk/SetTheory/

   This Web site declares that its aim is to cover a wide range of topics from ZFC Set Theory. Topics include: A brief History of Axiomatic Set Theory, Some Logical Paradoxes, The axioms of ZFC ,Order Relations, Equivalence Relations & Equipollence, Number Systems, Ordinals & Cardinals , and Glossary. Extensive hyperlinks are provided.

8. A History of Set Theory
   www-history.mcs.st-and.ac.uk/history/HistTopics/Beginnings_of_set_theory.html

   An article by J. J. O'Connor and E. F. Robertson. It discusses the history of Set Theory. It provides extensive hyperlinks. It has 25 references.

9. Boolean Logic
   www.howstuffworks.com/boolean.htm

   A tutorial in HTML format by Marshall Brain. Topics covered include: Introductory Information, Simple Gates, Flip-flops, and more.

10. Logic Gates and Boolean Logic
    educ.queensu.ca/~compsci/units/BoolLogic/titlepage.html

    A collection of annotated resources by Mark Mamo and Shane Bauman. Topics include:

    - Introduction to Boolean Logic
    - Black Box Circuits
    - Summary of Logic Gates
    - Sample Questions on Logic Gates, Circuits and Truth Tables
    - Discovering the Rules of Boolean Algebra
    - Simplifying Boolean Expressions

11. George Boole
    www.digitalcentury.com/encyclo/update/boole.html

    An article in HTML format by Natalie Voss. It provides information about George Boole and his work in addition to numerous related Web Links.

12. Max Black
    www-gap.dcs.st-and.ac.uk/~history/Mathematicians/Black.html

    An article in HTML format by J. J. O'Connor and E. F. Robertson with hyperlinks and seven references.

13. Lotfi Zadeh
    www.cs.berkeley.edu/~zadeh

    This site provides a bibliography of Lotfi Zadeh, who is considered the creator of fuzzy logic with his seminal paper published in 1965.

14. Fuzzy Logic Tutorial
    www.seattlerobotics.org/encoder/mar98/fuz/flindex.html

    A tutorial by Steven Kaehler that covers topics including:
    - Introduction to Fuzzy Logic
    - Why Use Fuzzy Logic
    - The Rule Matrix
    - Membership Functions
    - Putting It All Together
    - A Set of Sample Cases

15. Introduction to Fuzzy Logic
    www.dementia.org/~julied/logic/index.html

    A page maintained by Julie Grosman. It provides an introduction to fuzzy logic, its history, an overview of sets & operations, and its applications. It also has an interactive ten-question quiz.

16. Fuzzy Calculator
    http://home.planet.nl/~n.dubois/FzCalc/Fcalc.htm

    An interactive calculator for fuzzy operations by Nico du Bois. The calculator has a graphical interface through which one can define two three-element fuzzy sets by entering a number as an element and the corresponding membership function. The desired operation is performed by clicking on the related button of the calculator. The result is a three element set defined by the value of its elements and their corresponding membership values.

17. Fuzzy Sets and Systems Links
    www.abo.fi/~rfuller/fuzs.html

    A collection of Web links by Robert Fullér. It provides links to

    Personal Home Pages of Fuzzy Researchers, Who is Who in Fuzzy Database, Fuzzy-Mail Archives, Fuzzy Logic Sources of Information, Fuzzy logic at Webopedia, Computational, Conferences and Workshops on Fuzzy Systems, Professional Organizations, Journals & Books, Research groups, and more.

Fuzzy Relations

3.1 Introduction

Relationships among objects are fundamental to decision making and dynamic systems applications. A classical (crisp) relation represents the presence or absence of a connection or association among elements of two or more sets. Fuzzy relations, on the other hand, allow degrees of strength to such connections and associations.

This chapter gives a brief overview of classical relations: basic concepts, presentation, and operations to pave the way to the introduction of fuzzy relations along the same lines. The chapter ends with a discussion of fuzzy reasoning.

3.2 Classical Relations

A binary relation associates elements of only two sets to each other. A relation R from set A to set B is a subset of the Cartesian product of the two: $R \subseteq A \times B$. The relation can be represented by the notation aRb. Relations may also be defined among three sets (ternary relations), four sets (quaternary relations), five sets (quinary relations), and so on.

Presentation of relations

A relation can be defined by listing all the member pairs $(a,b) \in R$. It can also be expressed using a coordinate diagram, a matrix, a mapping diagram, or an arrow diagram.

Example

Let $A = \{\text{desk, bag, book}\}$, and $B = \{\text{wood, plastic, paper}\}$.

The Cartesian product of these two sets leads to
$A \times B = \{$(wood,desk), (wood,bag), (wood,book), (plastic,desk), (plastic,bag), (plastic,book), (paper,desk), (paper,bag), (paper,book)$\}$.
From this set one may select a subset such that
$R = \{$(wood,desk), (plastic,bag), ((paper,bag), (paper,book)$\}$.

The subset R can be represented using a coordinate diagram as shown in Figure 3.1.

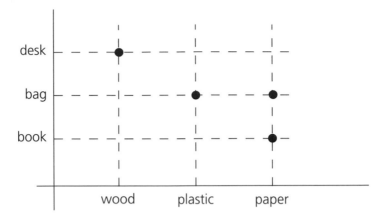

Figure 3.1: The coordinate diagram of a relation.

The relation could equivalently be represented using a matrix as shown below.

R	wood	plastic	paper
desk	1	0	0
bag	0	1	1
book	0	0	1

The mapping representation is yet another method of defining the relation between two sets. It is illustrated in Figure 3.2.

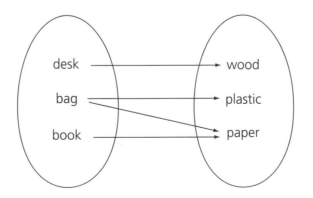

Figure 3.2: Mapping representation of a relation.

A binary relation in which each element from the first set is not mapped to more than one element in the second is referred to as a *function* and is expressed by

$R : A \rightarrow B$

Example

Figure 3.3 gives illustrations of $R : A \rightarrow B$.

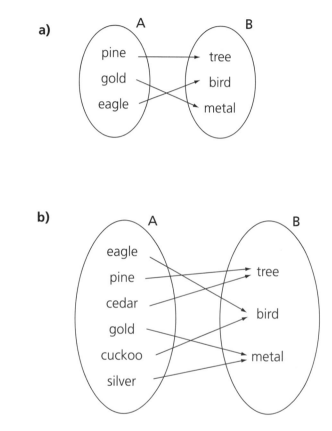

Figure 3.3: Illustrations of $R : A \rightarrow B$

The mapping can be arranged differently, leading to an arrow diagram.

Example

Let $A = \{1, 2, 3, 4, 5, 6\}$, and let R be a relation on A defined as
$R = \{(1,1), (1,2), (1,6), (3,2), (3,4), (4,5), (5,5), (5,6)\}$.
The relation can be represented by the arrow diagram shown in Figure 3.4.

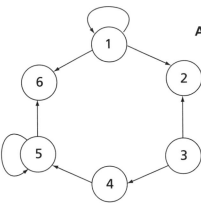

Figure 3.4:
An arrow diagram representation of *R*.

Operations on relations

a) Inverse of a relation

Let $R = \{(a,b)|(a,b) \in A \times B\}$, then the inverse, R^{-1}, is defined as

$R^{-1} = \{(b,a)|(a,b) \in R\}$. It follows that $bR^{-1}a$ iff aRb, and $(R^{-1})^{-1}$.

If $A = B$ and $R = R^{-1}$, the relation is referred to as *involuntary*, *self-inverse*, or *symmetric*.

If *R* is represented by a matrix M_R, then the matrix $M_{R^{-1}}$ representing R^{-1} is the transpose of $M_{R^{-1}} = M_R^T$. The transpose of a matrix is obtained by writing the rows, in order, as columns.

Example

Let $M_R = \begin{bmatrix} 1 & 1 & 0 \\ 0 & 0 & 0 \\ 1 & 0 & 0 \end{bmatrix}$, then $M_{R^{-1}} = \begin{bmatrix} 1 & 0 & 1 \\ 1 & 0 & 0 \\ 0 & 0 & 0 \end{bmatrix}$.

b) Composition

Composition is an operation on two compatible binary relations and results in a single binary relation. Two binary relations *P* and *Q* are compatible for composition iff

$P \subseteq A \times B$, and
$Q \subseteq B \times C$.

In other words, the second set in *P* must be the same as the first set in *Q*.

The composition of the two sets is denoted by $P \circ Q$. It can be determined graphically or using matrix representation where

$$M_{R \circ Q} = M_R M_Q$$

Some properties of the composition operation are summarized in Table 3.1.

Table 3.1: Some properties of the composition operation.

Associative	$(P \circ Q) \circ R = P \circ (Q \circ R)$
Not commutative	$P \circ Q \neq Q \circ P$
Inverse	$(P \circ Q)^{-1} = Q^{-1} \circ P^{-1}$

Example

Let $A = \{a_1, a_2, a_3\}$, $B = \{b_1, b_2, b_3\}$, and $C = \{c_1, c_2, c_3\}$.

Let $P = \{(a_1, b_1), (a_1, b_2), (a_2, b_3), (a_3, b_3)\}$, and

$Q = \{(b_1, c_1), (b_2, c_3), (b_3, c_1)\}$.

The relations P and Q are illustrated in Figure 3.5. From the illustration one may infer that:

$P \circ Q = \{(a_1, c_1), (a_1, c_3), (a_2, c_1), (a_3, c_1)\}$.

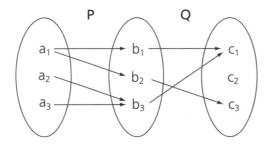

Figure 3.5: Illustration of the relation P and Q.

One can reach similar conclusions using the matrix representation as follows:

$$M_P = \begin{array}{c} \begin{array}{ccc} b_1 & b_2 & b_3 \end{array} \\ \begin{bmatrix} 1 & 1 & 0 \\ 0 & 0 & 1 \\ 0 & 0 & 1 \end{bmatrix} \begin{array}{c} a_1 \\ a_2 \\ a_3 \end{array} \end{array}, \text{ and}$$

$$M_Q = \begin{bmatrix} & c_1 & c_2 & c_3 & \\ 1 & 0 & 0 \\ 0 & 0 & 1 \\ 1 & 0 & 0 \end{bmatrix} \begin{matrix} b_1 \\ b_2 \\ b_3 \end{matrix}$$

$$\text{Then, } M_{P \circ Q} = \begin{bmatrix} & c_1 & c_2 & c_3 & \\ 1 & 0 & 1 \\ 1 & 0 & 0 \\ 1 & 0 & 0 \end{bmatrix} \begin{matrix} a_1 \\ a_2 \\ a_3 \end{matrix}$$

This matrix leads to

$$P \circ Q = \{(a_1,c_1), (a_1,c_3), (a_2,c_1), (a_3,c_1)\}.$$

as expected.

3.3 Classical Reasoning

In classical binary logic, reasoning is based on two complementary mechanisms: deduction (modus ponens) and induction (modus tollens). Deduction is used to obtain conclusions by means of forward inference and induction is used to deduce causes by means of backward inference. The two mechanisms are contrasted in Table 3.2. In that table A and B are crisp sets and the symbol $\rightarrow$ means implies.

Table 3.2: Deduction and Induction.

	Deduction	Induction
Rule	IF x is $A \rightarrow y$ is B	IF x is $A \rightarrow y$ is B
Premise	x is A	y is not B
Conclusion	y is B	x is not A

In other words, given the rule: :IF x is A, THEN y is B and the observation that "x is A", one concludes by deduction that : "y is B" . In a mathematical shorthand:

$$\left(p \wedge (p \rightarrow q)\right) \rightarrow q$$

Given the same rule but the observation that "y is not B", one concludes by induction that: "x is not A". In a mathematical shorthand:

$$\left(\bar{q} \wedge (p \rightarrow q)\right) \rightarrow \bar{p}$$

3.4 Fundamentals of Fuzzy Relations

Fuzzy relations are fuzzy sets defined on universal sets which are Cartesian products. They capture the strength of association among elements of two or more sets, not just whether such an association exists or not.

Example

Let $A = \{a_1, a_2, a_2\}$ and $b = \{b_1, b_2, b_3\}$.

Let R be a relation from A to B given by

$$R = 0.1/(a_1, b_3) + 0.5/(a_1, b_2) + 0.3/(a_2, b_1)$$
$$+ 0.4/(A_2, B_3) + 1.0/(A_3, B_3) + 0.1/(A_3, B_1)$$

The corresponding fuzzy matrix would be

$$M_R = \begin{array}{c} \begin{matrix} a_1 & a_2 & a_3 \end{matrix} \\ \begin{bmatrix} 0 & 0.3 & 0.1 \\ 0.5 & 0 & 0 \\ 0.1 & 0.4 & 1 \end{bmatrix} \begin{matrix} b_1 \\ b_2 \\ b_3 \end{matrix} \end{array}$$

The corresponding graph is shown in Figure 3.6.

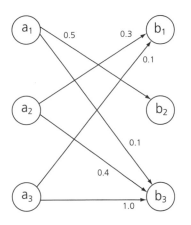

Figure 3.6:
Illustration of a fuzzy relation.

3.5 Operations on Binary Fuzzy Relations

Inverse of a fuzzy relation

The inverse of a fuzzy relation R on $A \times B$ is denoted by R^{-1}. It is a relation on $B \times A$ defined by $R^{-1}(b, a) = R(a, b)$ for all pairs $(b, a) \in B \times A$.

Example

$$\text{If } M_R = \begin{bmatrix} 0.3 & 1.0 & 0 \\ 1.0 & 0.5 & 0.3 \\ 0.0 & 0.0 & 0.8 \end{bmatrix} \text{, then}$$

$$M_{R^{-1}} = \begin{bmatrix} 0.3 & 1.0 & 0.0 \\ 1.0 & 0.5 & 0.0 \\ 0.0 & 0.3 & 0.8 \end{bmatrix}.$$

Composition of fuzzy relations

As explained in Section 3.2, the composition of two crisp binary relations P and Q requires that the relations be compatible—i.e., $A \times B$ and $B \times C$ share the set B. The composition $R = P \circ Q$ consists of pairs (a,b) from $A \times C$ that are connected through at least one element in B. In fuzzy relations, the connections have degrees of strength stemming from the fact P and Q are fuzzy sets.

There are several types of fuzzy relations compositions. The most common in engineering applications is the *max-min composition*. In this composition scheme the strength of the connection is determined by the smaller strength connection of the two in a chain that connects a to c. Among the chains that connect the two elements, the one with the largest strength is the one that is selected to characterize the relation. In mathematical shorthand, we can write

$$P \circ Q = \{\max \min[\mu_P (a,b), \mu_Q (b,c)]/(a,c)| \ a \in A, b \in B, c \in C\}$$

The properties of crisp relations given in Table 3.1 are valid for fuzzy relations composition as well.

Example

Let $A = \{a_1, a_2\}$,

$\quad B = \{b_1, b_2\}$, and

$\quad C = \{c_1, c_2\}$.

Let P be a relation from A to B defined by:

$$M_P = \begin{array}{c} \\ \\ \end{array} \begin{array}{cc} b_1 & b_2 \\ \begin{bmatrix} 0.4 & 0.2 \\ 0.8 & 0.3 \end{bmatrix} & \begin{array}{c} a_1 \\ a_2 \end{array} \end{array}$$

Let Q be a relation from B to C defined by:

$$M_Q = \begin{matrix} c_1 & c_2 \\ \begin{bmatrix} 0.3 & 0.1 \\ 0.4 & 0.9 \end{bmatrix} \begin{matrix} b_1 \\ b_2 \end{matrix} \end{matrix}$$

Then,

$$P \circ Q = \begin{bmatrix} 0.4 & 0.2 \\ 0.8 & 0.3 \end{bmatrix} \circ \begin{bmatrix} 0.3 & 0.1 \\ 0.4 & 0.9 \end{bmatrix}$$

$$= \begin{bmatrix} r_1 & r_2 \\ r_3 & r_4 \end{bmatrix}$$

where r_1, r_2, r_3, and r_3 are calculated as follows:

r_1 = max[min(0.4,0.3), min(0.2,0.4)]
   = max[0.3,0.2]
   = 0.3
r_2 = max[min(0.4,0.1), min(0.2,0.9)]
   = max[0.1,0.2]
   = 0.2
r_3 = max[min(0.8,0.3), min(0.3,0.4)]
   = max[0.3,0.3]
   = 0.3
r_4 = max[min(0.8,0.1), min(0.3,0.9)]
   = max[0.1,0.3]
   = 0.3

Hence, $M_{P \circ Q} = \begin{bmatrix} 0.3 & 0.2 \\ 0.3 & 0.3 \end{bmatrix}$

The same result could be obtained graphically by using an arrow diagram and applying the max-min rule of composition outlined earlier. This is illustrated in Figure 3.7.

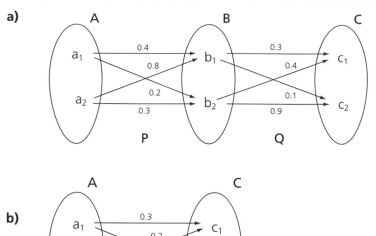

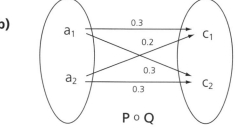

Figure 3.7: Illustration of the max-min composition.
a) The relations P and Q
b) The relation P ∘ Q

As mentioned previously, other types of composition of fuzzy relations could be defined. For example:

■ min-max composition (sometimes denoted by $P \square Q$) which is the dual of the max-min composition.

■ max-star composition where multiplication, summation, or some other binary operation is denoted by ★ in place of the min operation in the max-min composition.

Two special cases of max-star composition are the max-product (max-prod) and the max-average (max-av) compositions.

3.6 Types of Fuzzy Relations

If R is a fuzzy relation on $A \times A$, and $x \in A$, $y \in A$, and $z \in A$, then the relation R can be classified as summarized in Table 3.3.

Table 3.3: Summary of fuzzy relations types.

Reflexive	$\mu_R\,(x,x) = 1$ and $\mu_R\,(x,y) < 1,\, x \neq y$
Anti-reflexive	$\mu_R\,(x,x) = 0$ and $\mu_R\,(x,y) < 1,\, x \neq y$
Symmetric	$\mu_R\,(x,y) = \mu_R\,(y,x)$
Anti-symmetric	$\mu_R\,(x,y) \neq \mu_R\,(y,x)$
max-min transitive	$\mu_R\,(x,z) \geq \max\min[\mu\,(x,y),\, \mu\,(x,z)]$
Identity relation	Diagonal elements of M_R are all ones
Zero relation	All elements of M_R are zeros
Universal	All elements of M_R are ones

Examples

The following are examples of M_R matrices for reflexive, anti-reflexive, symmetric and antisymmetric relations, respectively:

$$\begin{bmatrix} 1.0 & 0.2 & 0.7 \\ 0.3 & 1.0 & 0.5 \\ 0.4 & 0.6 & 1.0 \end{bmatrix}, \begin{bmatrix} 0.0 & 0.2 & 0.8 \\ 0.5 & 0.0 & 0.7 \\ 0.2 & 0.3 & 0.0 \end{bmatrix}, \begin{bmatrix} 0.3 & 0.7 & 0.5 \\ 0.7 & 0.8 & 0.2 \\ 0.5 & 0.2 & 0.4 \end{bmatrix}, \text{ and}$$

$$\begin{bmatrix} 0.3 & 0.5 & 0.8 \\ 0.2 & 1.0 & 0.5 \\ 0.4 & 0.6 & 0.2 \end{bmatrix}.$$

3.7 Fuzzy Reasoning

Fuzzy reasoning is based on inference rules of the form

> IF <premise>, THEN <consequence>

as is the case in classical logic, but fuzzy sets, rather than crisp sets, are used. Fuzzy sets define linguistic variables and hence fuzzy inference rules can model a system linguistically. Fuzzy algorithms are mathematically equivalent to fuzzy relations and fuzzy inference is equivalent to fuzzy composition.

There are numerous ways that have been put forward to express an inference rule. A direct, simple inference rule takes the form:

> IF x is A, THEN y is B

where A and B are fuzzy sets.

If the number of rules is large it becomes more convenient to employ a fuzzy relations approach. The IF/THEN rules are converted to fuzzy relations, then fuzzy composition is used to infer conclusions. The conversion from IF/THEN rules to fuzzy relations could be defined in more than one way. A simple method is given by:

$$R = A \rightarrow B = A \times B$$

An inference rule could have more than one proposition. For example, a rule of inference with two propositions would take the form:

if x is A, and y is B, then z is C

where A, B, and C are fuzzy subsets of X, Y, and Z, respectively.

The rule may be written as:

A and $B \rightarrow C$

Since A and B are subsets of different sets X and Y, respectively, it would be inaccurate to write $A \cap B \rightarrow C$. The situation is illustrated in Figure 3.8, from which one concludes that

"x is A and y is B" is $\Leftrightarrow (x, y)$ is $(A \times Y) \cap (X \times B)$

$$(x, y) \, is \, A \times B$$

Hence, the proposition "A and B" can be expressed as

A and $B = A \times B$

Then the relation R can be expressed as

$$R = A \text{ and } B \rightarrow C$$
$$= A \times B \times C$$

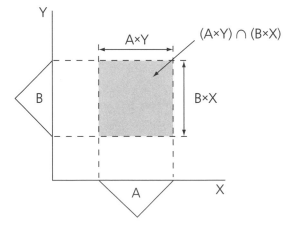

Figure 3.8: $A \cap B$ with A and B being subsets of different sets X and Y.

A fuzzy algorithm has several rules, such as

Rule 1:   IF x is A_1, THEN y is B_1

Rule 2:   IF x is A_2, THEN y is B_2

.

.

.

Rule n: IF x is A_n, THEN y is B_n

The rules can be also written as

$A_1 \rightarrow B_1$

$A_2 \rightarrow B_2$

.

.

.

$A_n \rightarrow B_n$

This n-rule system can be converted to n relations: $R_1, R_2,...,R_n$. These relations can be combined into one relation, R, using fuzzy intersection operations or fuzzy union operations depending on how the rules are perceived to be connected.

$R = R_1 \cup R_2 ... \cup R_n$, or

$R = R_1 \cap R_2 ... \cap R_n$

The difference in the way the rules are perceived to be related would obviously lead to different results.

Concluding Remarks

A crisp relation represents the presence or absence of association or interaction between the elements of two or more sets. As the concept of crisp sets was generalized to allow degrees of membership, so was the concept of relations. Fuzzy relations allow degrees of association or interaction between elements to better reflect one's way of thinking and decision making. They may be thought of as fuzzy sets defined over multidimensional universes of discourse. Fuzzy relations play a central role in fuzzy algorithms, which are collections of fuzzy rules of the form of IF/THEN. Such rules are nothing other than fuzzy relations in disguise. Embedded fuzzy system applications are based on hardware and software implementations of fuzzy algorithms using embedded technology approaches.

Bibliography

1. M. Benedicty and R. F. Sledge, *Discrete Mathematical Structures*, HBJ, Toronto, 1987.

2. E. Cox, *The Fuzzy Systems Handbook*, AP Professional, Boston, USA, 1994.

3. A. Kaufman and M. M. Gupta, *Introduction to Fuzzy Arithmetic*, Van Nostrand Reinhald, New York, USA, 1991.

4. G. J. Klir and T. A. Folger, *Fuzzy Sets, Uncertainty, and Information*, Prentice Hall, London, UK, 1998.

5. G. J. Klir, U. H. St. Clair, and B. Yuan, *Fuzzy Set Theory*, Prentice Hall, Upper Saddle River, NJ, USA, 1997.

6. E. H. Mamdani, "Application of Fuzzy Algorithms for Control of a Simple Dynamic Plant", *Proceedings of IEEE*, 121, 12, 1585–1588, 1974.

7. M. Sugeno, "An Introductory Survey on Fuzzy Control", *Information Sciences*, 36, 59–83, 1985.

8. T. Tanaka, *An Introduction to Fuzzy Logic for Practical Applications*, Springer, New York, USA, 1997.

9. T. Terano, K. Asai, and M. Sugeno, *Fuzzy Systems Theory and Applications*, Academic Press, Inc., Boston, USA, 1992.

Web Resources

1. On Fuzzy Relations, Metrics and Cluster Analysis
    http://dmi.uib.es/people/valverde/gran1/GRAN1.html

    An article in HTML format by J. Jacas and L. Valverde. It gives an introduction to fuzzy logic and discusses: Building fuzzy transitive relations, Generators, m-cluster coverages. It provides 20 references.

2. Modelling Linguistic Expressions Using Fuzzy Relations
    www.scch.at/servlet/resource.ResourceLoader?id=5

    A 15-page technical report with 13 references available in PDF format. It is authored by M. De Cock, U. Bodenhofer, and E. Kerre, and appeared in The Proceedings of the 6th International Conference on Soft Computing, Iizuka, Japan, October 1-4, 2000, pp. 353–360.

3. Numerical Representation of Transitive Fuzzy Relations
    http://online.sfsu.edu/~sergei/pr-05.pdf

    A 12-page paper with 12 references by S. Ovchinnikov, Mathematics Department, San Francisco Stat University. It investigates numerical representations of fuzzy transitive relations.

4. Fuzzy Researchers
    www.abo.fi/~rfuller/persons.html

    This site provides a list of home pages of fuzzy logic researchers. It is compiled by Robert Fuller, Åbi Akademi, Finland.

CHAPTER 4

Embedded Fuzzy Applications

4.1 Introduction

As outlined in Chapter 1, there are various definitions of an embedded system. An embedded system involves computing, but not as its main purpose. Fuzzy logic has a wide range of applications in such systems. The common thread that links all these applications is that an expert (a human operator) can, at least in principle, perform the task required. However, the mathematical model is too complicated to be of practical use, and therefore a linguistic model becomes advantageous. In some cases where mathematical models do exist and lead to satisfactory results, fuzzy logic may still be used either to improve or simplify the process or to compare with a benchmark. There are numerous embedded applications of fuzzy logic such as fuzzy control, fuzzy signal processing, and fuzzy image processing. The discussion here will concentrate on fuzzy control. Analysis and design of conventional control systems are overviewed first, to be contrasted afterwards with fuzzy logic control analysis and design.

4.2 Conventional Control Systems

4.2.1 Description

A typical block diagram of a feedback control system is shown in Figure 4.1.

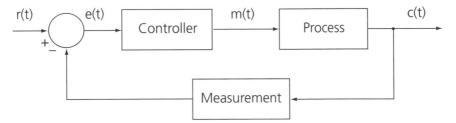

Figure 4.1: A typical block diagram of a feedback control system.

The system consists of three main blocks: the process to be controlled, measurement, and the controller. The output of the process is measured and converted to an equivalent electrical signal using suitable sensor circuitry. The signal that represents the output, $c(t)$, is then compared with an input reference signal, $r(t)$, resulting in an error signal, $e(t)$. This error signal actuates the controller to generate the control action signal, $m(t)$.

The signals mentioned in the above descriptions are expressed as functions of time—i.e., the description is in the *time domain*. The system is described using a differential equation with the time-derivative operator d/dt being the basic operator used. The system could have been equivalently described in the *frequency domain*, or the *s-domain*. The Laplace operator, s, is the basic operator used in the s-domain description of the system. It is equivalent to the time-derivative operator in the time domain (with the initial conditions being zero). The derivative operator is also equivalent to the $j\omega$ operator for sinusoidal signals, which is used in the frequency domain description of a system. The ratio of the output to the input signals expressed in the s-domain is the transfer function of the system, $G(s)$, with s being, in general, a complex number expressed by $s = \sigma + j\omega$. The basic forms of the transfer function include:

- Gain, $G(s) = K$

- Differentiator, first-order lead, $G(s) = \tau s$

- Integrator, first-order lag, $G(s) = (\tau s)^{-1}$

- Second-order lag, $G(s) = (s^2 + 2\zeta\omega_n s + \omega_n^2)^{-1}$

where K is a constant (gain factor), τ is the time-constant, ζ is the damping factor, and ω_n is the natural frequency.

The transfer function of a system is, in general, composed of such basic transfer functions. It can then be expressed in one of two equivalent forms:

- Pole-zero form:

$$G(s) = \frac{K(s+z_1)(s+z_2)(s+z_3)...}{(s+p_1)(s+p_2)(s+p_3)...}$$

  where $-z_1, -z_2, -z_3,...$ and $-p_1, -p_2, -p_3,...$ are the zeros and poles of the system, respectively.

- Time-Constant form:

$$G(s) = \frac{K(\tau_1 s+1)(\tau_2 s+1)(\tau_3 s+1)...}{(\tau_a s+1)(\tau_b s+1)(\tau_c s+1)...}$$

4.2.2 Analysis

The objectives of analyzing a given feedback system are mainly to determine:

- the stability of the system and its extent
- the transient response
- the steady-state response

A system is said to be stable when its output does not increase without bound or oscillate with time. The steady-state and transient response parameters are defined in Figure 4.2.

To analyze a feedback control system one needs to:

- Determine a mathematical model for each of the system building units. Transfer functions are the most convenient to use for that purpose.
- Represent the system using a block diagram and determine its overall transfer function.
- Determine the characteristics by means of a step-response, pole-zero map, root-locus, Bode plot, Nyquist diagram, or Nicols Chart.

Simply from the transfer function point of view, the stability of a single-input, single-output system can be determined based on the location of the poles in the complex plane (*s*–plane): the left half of the complex plane is the stable region.

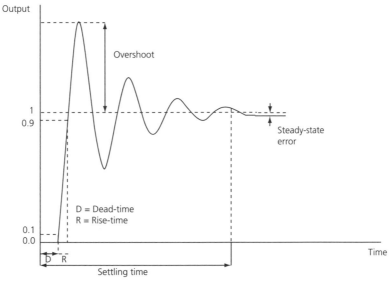

Dead time: the time elapsed between the application of an input signal and the start of an output signal.
Rise time: the time the output takes to increase from 10% to 90% of the final value.
Overshoot: the maximum difference between the transient and steady-state values.
Settling time: the time for the response to become within 5% of the final value.
Steady-state error: the value of $e(t)$ as $t \rightarrow \infty$.

Figure 4.2: A possible step response of a stable system with the steady-state and transient response parameters defined.

4.2.3 Design

The objective of the design of a feedback control system is to meet a set of given performance specifications. They are usually given in the time-domain or the frequency domain.

The time-domain specifications have two components:

- Transient performance, which could be given in terms of overshoot percentage, rise time, settling time, and the predominant time constant.

- Steady-state performance, which is given in terms of steady-state error.

The frequency-domain specifications are usually given in terms of gain and phase margins, cut-off rate, and resonance peak and frequency.

4.2.4 PID Control

A commonly used control scheme is known as the three-term PID controller, where PID stands for Proportional Integral Derivative. It is also used as a benchmark against which any other control scheme is compared. The concept of a PID controller is illustrated in Figure 4.3.

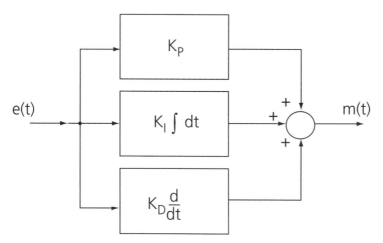

Figure 4.3: PID controller concept.

The control action signal of a PID controller is composed of three components:

- Proportional action, $p(t)$, where the signal is proportional to the error signal at the present moment.

$$p(t) = K_p e(t)$$

- Integral action, $i(t)$, where the signal is proportional to the commutative values of the error signal up to the present moment.

$$i(t) = K_i \int_0^t e(t) dt$$

- Derivative action, $d(t)$, which is proportional to the rate of change of the error signal at the present moment.

$$d(t) = K_d \frac{d}{dt} e(t)$$

where K_p, K_i, K_d and are constants.

The overall control action, $m(t)$, can be expressed as:

$$m(t) = K_p e(t) + K_i \int_0^t e(t) dt + K_d \frac{d}{dt} e(t)$$

In the s-domain, this can be expressed as:

$$M(s) = \left(K_p + \frac{K_1}{s} + K_D s \right) \cdot E(s)$$

As K_p is increased, the system speed increases (with a tendency to overshoot), and the steady-state error decreases, but is not eliminated. As K_d is increased, the damping factor increases, thereby reducing the overshoot. The derivative control is susceptible to noise and it is never used alone. As K_i is increased, the steady-state error goes to zero and the system tends towards instability. Integral control is also never used alone.

4.3 Fuzzy Logic Controller (FLC)

4.3.1 Description

The block diagram shown in Figure 4.1 is still applicable in fuzzy applications, but the controller is a fuzzy controller rather than a PID controller. Figure 4.4 shows the basic structure of a fuzzy logic controller.

The main building units of an FLC are a fuzzification unit, a fuzzy logic reasoning unit, a knowledge base, and a defuzzification unit. Defuzzification is the process of converting inferred fuzzy control actions into a crisp control action.

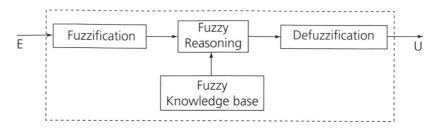

Figure 4.4: Basic structure of a fuzzy logic controller.

The fuzzy knowledge-base has a rule-base that maps a fuzzy input variable, E, into a fuzzy output, U. This can be expressed by a linguistic statement such as:

$E \rightarrow U$ (condition E implies condition U)

which may be written as:

IF E THEN U.

There is an equivalency between the above expression and the relation obtained by Cartesian multiplication, as explained in Chapter 3—i.e.,

$R = E \times U \equiv$ IF E THEN U.

The fuzzy knowledge-base also has a database defining the variables. A fuzzy variable is defined by a fuzzy set, which in turn is defined by a membership function.

Fuzzy reasoning is used to infer the output contributed from each rule. The fuzzy outputs reached from each rule are aggregated and defuzzified to generate a crisp output.

The essence of a fuzzy logic controller is thus based on a linguistic model (rule-base and the defined membership functions) as opposed to a mathematical model, as is the case with a PID controller. Fuzzy logic controllers are used to reduce the development time or to improve the performance of an existing PID controller. In the case of highly complex systems, fuzzy logic could be the only solution.

4.3.2 Design

In the design of an FLC system it is assumed that:

- A solution exists.

- The input and output variables can be observed and measured.

- An adequate solution (not necessarily an optimum one) is acceptable.

- A linguistic model can be created based on the knowledge of a human expert.

In order to model a system linguistically, one needs to:

- Identify the input and output variables of the process to be controlled (the plant). For example: speed, temperature, humidity, etc.

- Define subsets that cover the universe of discourse of each variable and assign a linguistic label to each one. For example, the linguistic variable *speed* may be defined as three fuzzy subsets: *slow, medium*, and *fast* as shown in Figure 2.9.

- Form a rule-base by assigning relationships between inputs and outputs.

- Determine a fuzzification method.

- Determine a defuzzification method to be used to generate a crisp output from the fuzzy outputs generated from the rule-base. The section to follow elaborates on the process of defuzzification and its numerous methods.

4.3.3 Defuzzification

For a given input, several IF/THEN rules could be launched at the same time. Each rule would have a different strength, because a given input may belong to more than one fuzzy set, but with different membership values. For example, an input temperature of 80 °C may belong to the fuzzy subset *very_high* with $\mu = 0.8$ and to the fuzzy subset *medium* with $\mu = 0.3$. Thus, when this temperature occurs two, rules will fire:

IF very_high THEN action-1

IF medium THEN action-2

If action-1 is defined by fuzzy set F_1 and action-2 is defined by fuzzy set F_2, then the two sets are aggregated (commonly using the UNION operation) leading to the fuzzy set F, as illustrated in Figure 4.5.

In general, the output of the fuzzy reasoning would involve more than two fuzzy sets; therefore, one can write:

$$F = \bigcup_{i=1}^{k} F_i$$

Assuming the support of F is $X = \{x_1, x_2, x_3, \ldots\}$ then for $x_i \in X$, $F(x_i) = w_i$ indicates the degree to which each is suggested by the rule-base as a good output for the given input. The defuzzification operation is applied on F to determine the best crisp output.

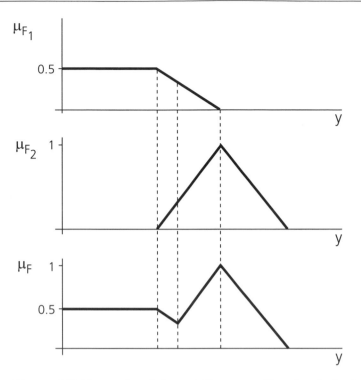

Figure 4.5: Two rules firing at the same time leads to fuzzy sets that could be aggregated and result in a fuzzy set *F*.

Numerous defuzzification methods have been suggested in the literature; however, sometimes different authors name the same method differently. No method has proved to be always more advantageous than the others. The selection of which method to use depends primarily on the experience of the designer. One may take into consideration the computational complexity involved, applicability to all situations considered, and plausibility of the results from an engineering point of view. Defuzzification methods include the following:

Centroid method

This method is also known as the *center of mass*, or *center of gravity* method. It is probably the most commonly used defuzzification method. The defuzzified output, x^* is defined by

$$x^* = \frac{\int \mu_F(x) \cdot x\,dx}{\int \mu_F(x)\,dx}$$

where the symbol $\int$ denotes algebraic integration.

Center of largest area

This method is applicable when the output consists of at least two convex fuzzy subsets which do not overlap. The result is biased towards a side of one membership function.

First maxima

This method is applicable when the output is peaked; the smallest value of the domain with maximum membership is selected.

Max-membership

This method is also known as the *height* method. The output is defined by:

$$\mu_F\left(x^*\right) \ge \mu_F\left(x\right), \text{ for all } x \in X$$

Weighted average method

This method is valid for symmetrical output membership functions. Each membership function is weighted by its maximum membership value. The output is defined by:

$$x^* = \frac{\sum \mu_F\left(\overline{x}_i\right) \cdot \overline{x}_i}{\sum \mu_F\left(\overline{x}_i\right)}$$

where $\sum$ denotes algebraic summation, and $\overline{x}_i$ is the maximum of the ith membership function.

The method is illustrated in Figure 4.6 where two fuzzy subsets are considered. In that figure, the defuzzified output would be

$$x^* = \frac{\left(0.4\right)a + \left(0.8\right)b}{0.4 + 0.8}$$

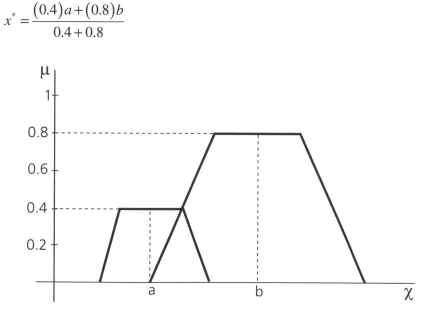

Figure 4.6: Two symmetrical membership functions.

Mean-max membership

The method is also known as the *middle of the maxima*. The output is defined by:

$$x^* = \frac{\sum_{i=1}^{n} \overline{x}_i}{n}$$

In Figure 4.6, the defuzzified output would be:

$$x^* = \frac{a+b}{2}$$

Center of sums

This method uses the algebraic sum of the individual fuzzy subsets instead of their union. Although the calculations become faster, this method leads to adding the intersecting areas twice.

4.3.4 Analysis

As discussed earlier, fuzzy logic is particularly useful in the control of highly nonlinear processes difficult to control by conventional methods. The central idea is to use the knowledge of an expert (human operator) to create a linguistic model. However, conventional analysis techniques of control systems are based on mathematical models and hence cannot be used for the analysis of a fuzzy logic control system. One may argue that if the model is based on the knowledge of an expert then the analysis is not needed. However, one must realize that human knowledge could be incomplete. In addition, fidelity of the mapping using rule-base, inference rules, and defuzzification methods needs to be assured. Extensive simulations, in general, would be required to establish confidence in the reliability of the system. One should examine the static properties of a designed FLC system and observe aspects such as:

- The consistency of the IF/THEN rules. Inconsistencies could occur due to an attempt to accommodate contradictory design criteria or simply due to faulty formulation of the rules.

- The existence of enough rules to assure smooth control action, but not too many to slow the operation.

- A proper control action should be inferable for every state of the process.

- Fuzzy sets are properly assigned in the defined universe of discourse. Overlapping of neighboring membership functions ensures completeness and robustness, but may lead to distortions in the output.

- The defuzzification method used is suitable to ensure correct and smooth control action.

As for the dynamic properties, numerous techniques have been put forward for the analysis of the stability of an FLC system. They can be categorized into two groups:

- The first group consists of methods that require a mathematical model of the process to be controlled (the plant). This category includes the cases of linear and nonlinear plants. The describing function technique and the circle criterion are example of the first. The invariance principle and Lyapunov function are examples of the second.

- The second group consists of methods that do not require a mathematical model. They include phase plane trajectory, energetistic stability, and linguistic stability methods.

Energetistic Stability Criterion

The energetistic stability criterion involves time-consuming calculations; however, it provides an intuitive method to analyze the stability of a fuzzy dynamic system. It is based on the general physical law that any dynamic system is stable if its total energy decreases with time until it reaches a steady state. Various authors have suggested algorithms based on this idea. The approach of Kiszka et al. is summarized here. It starts with the basic equation of a simple dynamic system,

$$X_{k+1} = X_k \circ U_k \circ R, k = 1, 2, 3, \ldots$$

where X_k and X_{k+1} are fuzzy sets at the kth and $(k+1)^{th}$ time instants, respectively, and R is a fuzzy relation describing the fuzzy control system.

If there is no input, the system is said to be free or unforced. This occurs when U_k = zero for all values of k, where *zero* is a fuzzy singleton as illustrated in Figure 4.7 and is described by

$$\mu_{U_k}(u) = \text{zero} = \begin{cases} 1 & u = u_1(k) \\ 0 & \text{otherwise} \end{cases} \qquad \text{for all values of } k$$

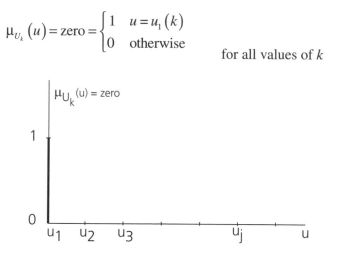

Figure 4.7: Illustration of the fuzzy singleton, U_k = zero.

Let , $P = U_k \circ R$ then $X_{k+1} = X_k \circ P$, $\qquad k = 0,1,2,3,\ldots$

A state is an equilibrium state if $X_{k+1} = X_k = X_S$ for all k. This leads to:

$$X_S = X_S \circ P$$

The energy of a fuzzy relation P is defined by Kiszca et al. by:

$$E(P) = \frac{1}{n \cdot m} \sum_{i=1}^{n} \sum_{j=1}^{m} w(x_i, y_j) \cdot f\left(\mu_P(x_i, y_j)\right)$$

where $w(x_i, y_j)$ is the Cartesian product of the ith element in the support of the fuzzy set X in the universe of discourse W_X, the jth element in the support of fuzzy set Y in the universe of discourse W_Y, $f\left(\mu_P(x_i, y_j)\right)$, is the corresponding value of the membership in the normalized fuzzy relation matrix M_P, and n and m are the cardinalities of the support sets for fuzzy sets X and Y, respectively.

Example

Let $W_X = [2,4,6]$, $W_Y = [1,3,5]$, and P is a relation from X to Y defined by

$$M_P = \begin{bmatrix} 0.2 & 0.3 & 0.1 \\ 0.1 & 0.4 & 0.3 \\ 0.5 & 0.1 & 0.2 \end{bmatrix}$$

Then, the energy would be:

$$E(P) = \frac{1}{(3)(3)}\{2 \times 1 \times 0.2 + 2 \times 3 \times 0.1 + 2 \times 5 \times 0.5$$

$$+ 4 \times 1 \times 0.3 + 4 \times 3 \times 0.4 + 4 \times 5 \times 0.1$$

$$+ 6 \times 1 \times 0.1 + 6 \times 3 \times 0.3 + 6 \times 5 \times 0.2\}$$

$$= 2.89$$

The energy expression for a free fuzzy dynamic system would be:

$$E(X_{k+1}) = E(X_k \circ P)$$

But, $X_1 = X_0 \circ P$,

$$X_2 = X_1 \circ P = X_0 \circ P \circ P = X_0 \circ P^2,$$

.

.

.

$$X_k = X_0 \circ P^k$$

where X_0 is the initial fuzzy system state and $P^k = P \circ P \circ P \ldots P$ (k times).

Then, the energy expression can be written as:

$$E(X_{k+1}) = E(X_0 \circ P^{k+1})$$

The change in energy, ΔE, between two consecutive states would be:

$$\Delta E = E(X_0 \circ P^k) - E(X_0 \circ P^{k-1})$$

Since $E(X_0)$ is a constant, ΔE does not depend on the initial conditions. Therefore, the rational *characteristic energy function*, $\Delta E_C(P,k)$, can be expressed as

$$\Delta E_C(P,k) = E(P^k) - E(P^{k-1})$$

This expression could be used to determine the stability as follows:

a) The system is stable if $\Delta E_C(P,k) \le 0$ for $k \to \infty$.

b) The system is unstable if $\Delta E_C(P,k) \ge 0$ for $k \to \infty$.

c) The system is oscillatory with a period T, if

$$\left| \Delta E_C(P,k) \right| = \left| \Delta E_C(P,k+T) \right| \text{ for } k \to \infty.$$

The following is a simple illustrative example of the energetistic stability criterion as put forward by Kiszka et al.

Example

Let the universe of discourse be $X = Y = [1,2]$. Determine the stability of the system if the fuzzy relation P is given by:

a) $P = \begin{bmatrix} 0.1 & 0.5 \\ 0.3 & 0.2 \end{bmatrix}$

b) $P = \begin{bmatrix} 0.1 & 0.3 \\ 0.7 & 1.0 \end{bmatrix}$

To determine the stability of a system, one starts by calculating several consecutive values of $E(P^k)$ using the max-min composition rule discussed in Chapter 3. This gives

$$\begin{bmatrix} a_{11} & a_{12} \\ a_{21} & a_{22} \end{bmatrix} \circ \begin{bmatrix} b_{11} & b_{12} \\ b_{21} & b_{22} \end{bmatrix} = \begin{bmatrix} c_{11} & c_{12} \\ c_{21} & c_{22} \end{bmatrix}$$

with

$$c_{11} = \max\left[\min(a_{11},b_{11}), \min(a_{12},b_{21}) \right]$$
$$c_{12} = \max\left[\min(a_{11},b_{12}), \min(a_{12},b_{22}) \right]$$
$$c_{21} = \max\left[\min(a_{21},b_{11}), \min(a_{22},b_{21}) \right]$$
$$c_{22} = \max\left[\min(a_{21},b_{12}), \min(a_{22},b_{22}) \right]$$

Then for the first system:

$$P^2 = P \circ P = \begin{bmatrix} 0.1 & 0.5 \\ 0.3 & 0.2 \end{bmatrix} \circ \begin{bmatrix} 0.1 & 0.5 \\ 0.3 & 0.2 \end{bmatrix}$$

$$= \begin{bmatrix} 0.3 & 0.2 \\ 0.2 & 0.3 \end{bmatrix}$$

$$P^3 = P^2 \circ P = \begin{bmatrix} 0.3 & 0.2 \\ 0.2 & 0.3 \end{bmatrix} \circ \begin{bmatrix} 0.1 & 0.5 \\ 0.3 & 0.2 \end{bmatrix}$$

$$= \begin{bmatrix} 0.2 & 0.3 \\ 0.3 & 0.2 \end{bmatrix}$$

$$P^4 = P^3 \circ P = \begin{bmatrix} 0.2 & 0.3 \\ 0.3 & 0.2 \end{bmatrix} \circ \begin{bmatrix} 0.1 & 0.5 \\ 0.3 & 0.2 \end{bmatrix}$$

$$= \begin{bmatrix} 0.3 & 0.2 \\ 0.2 & 0.3 \end{bmatrix}$$

One can then use the definition of energy

$$E(P) = \frac{1}{n \cdot m} \sum_{i=1}^{n} \sum_{j=1}^{m} w(x_i, y_j) \cdot f\left(\mu_P(x_i, y_j)\right) \text{ to calculate}$$

$$E(P), E(P^2), E(P^3), E(P^4)\ldots$$

$$E(P) = \frac{1}{2 \times 2} \{1 \times 1 \times 0.1 + 1 \times 2 \times 0.5 + 2 \times 1 \times 0.3 + 2 \times 2 \times 0.2\}$$

$$= \frac{1}{4}(2.5)$$

Similar calculations yield:

$$E(P^2) = \frac{1}{4}(2.3), \ E(P^3) = \frac{1}{4}(2.2), \text{ and } E(P^4) = \frac{1}{4}(2.2).$$

One may conclude that the system is oscillatory.

For the second system:

$$P^2 = \begin{bmatrix} 0.1 & 0.3 \\ 0.7 & 1.0 \end{bmatrix} \circ \begin{bmatrix} 0.1 & 0.3 \\ 0.7 & 1.0 \end{bmatrix} = \begin{bmatrix} 0.3 & 0.3 \\ 0.7 & 1.0 \end{bmatrix}$$

$$P^3 = \begin{bmatrix} 0.3 & 0.3 \\ 0.7 & 1.0 \end{bmatrix} \circ \begin{bmatrix} 0.1 & 0.3 \\ 0.7 & 1.0 \end{bmatrix} = \begin{bmatrix} 0.3 & 0.3 \\ 0.7 & 1.0 \end{bmatrix}$$

$$P^4 = P^5 = \begin{bmatrix} 0.3 & 0.3 \\ 0.7 & 1.0 \end{bmatrix}$$

One may then conclude that this system is stable.

4.4 Simplified Examples of Applications

Two examples are given here of fuzzy logic control in embedded systems: a washing machine and a vacuum cleaner. The objective is to illustrate how to start thinking about designing a practical fuzzy logic system, the core of which is the definition of the inputs, outputs and fuzzy control rules. The two examples presented do not require extensive knowledge. Hence, the rules, the fuzzy subsets, and the corresponding results of simulation are easily understood. By relating them to each other and reflecting on how to enhance their operation, they should provide an insight into the operation of more complex systems that may require specialized knowledge. No calculation details are given here. Usually such design and analysis is carried out with the help of a software tool. Some of the software tools available will be discussed in Chapter 9. Web links to some fuzzy embedded systems, including washing machines, vacuum cleaners, and car anti-lock brake systems are provided at the end of the chapter.

4.4.1 Washing machine

There are several washing machines on the market that use fuzzy logic. The discussion here, however, does not refer to any particular washing machine; it is rather concerned with a hypothetical washing machine. The objective is to provide a simple example of the initial stages of designing a fuzzy control for a washing machine.

Ease of use is a desirable feature, along with ability to set the washing parameters based on the laundry load characteristics. The characteristics of the laundry load (inputs) include: the actual weight, fabric types, and amount of dirt. The washing parameters (outputs) include: amount of detergent, washing time, agitation, water level, and temperature. Controlling these parameters could lead to a cleaner laundry, conserve water, save detergent, electricity, time, and money.

The design of a machine to meet such specifications could be a demanding task. It is obvious that there is no simple mathematical model that could be of practical use to relate the inputs to the outputs, but a knowledgeable operator could do the task manually. This is where fuzzy logic comes in. A rule-base could be created based on the knowledge of the operator to control the process. Analysis would still be necessary to assure that plausible results are reached.

Consider, for simplicity, a machine with two inputs and one output, the inputs being:

- The dirtiness of the load as measured by the opacity of the washing water using an optical sensor system.

- The weight of the laundry load as measured by a pressure sensor system.

The output is the amount of detergent dispensed.

The dirtiness is defined in the range from 0 to 100, by defined fuzzy subsets: *Almost_Clean, Dirty, Soiled,* and *Filthy* as shown in Figure 4.8. The weight of the laundry is defined in the range from 0 to 100 by fuzzy sets: *Very_Light, Light, Heavy,* and *Very_Heavy* as shown in Figure 4.9. The output, for simplicity, is defined by the singleton subsets shown in Figure 4.10.

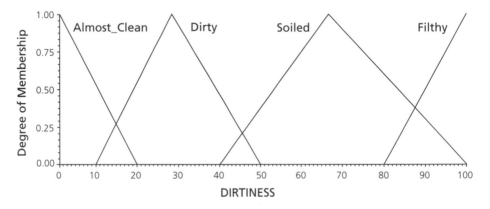

Figure 4.8: Subsets defining *Dirtiness*.

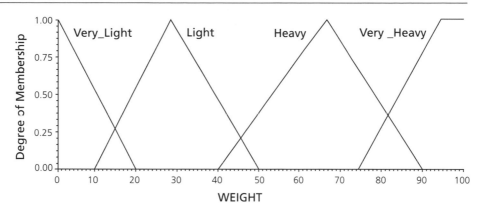

Figure 4.9: Subsets defining *Weight*.

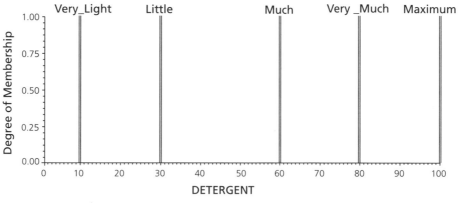

Figure 4.10: The *Detergent* subsets.

Suggested control rules are given below, and summarized in Table 4.1.

```
if DIRTINESS is ALMOST_CLEAN and WEIGHT is VERY_LIGHT then DETERGENT is VERY_LITTLE
if DIRTINESS is DIRTY and WEIGHT is VERY_LIGHT then DETERGENT is LITTLE
if DIRTINESS is SOILED and WEIGHT is VERY_LIGHT then DETERGENT is MUCH
if DIRTINESS is FILTHY and WEIGHT is VERY_LIGHT then DETERGENT is VERY MUCH
if DIRTINESS is ALMOST_CLEAN and WEIGHT is LIGHT then DETERGENT is VERY_LITTLE
if DIRTINESS is DIRTY and WEIGHT is LIGHT then DETERGENT is LITTLE
if DIRTINESS is SOILED and WEIGHT is LIGHT then DETERGENT is MUCH
if DIRTINESS is FILTHY and WEIGHT is LIGHT then DETERGENT is MUCH
if DIRTINESS is ALMOST_CLEAN and WEIGHT is HEAVY then DETERGENT is MUCH
if DIRTINESS is DIRTY and WEIGHT is HEAVY then DETERGENT is MUCH
if DIRTINESS is SOILED and WEIGHT is VERY_HEAVY then DETERGENT is VERY_MUCH
if DIRTINESS is FILTHY and WEIGHT is HEAVY then DETERGENT is MAXIMUM
if DIRTINESS is ALMOST_CLEAN and WEIGHT is VERY_HEAVY then DETERGENT is MUCH
if DIRTINESS is DIRTY and WEIGHT is VERY_HEAVY then DETERGENT is VERY_MUCH
if DIRTINESS is SOILED and WEIGHT is VERY_HEAVY then DETERGENT is MAXIMUM
if DIRTINESS is FILTHY and WEIGHT is VERY_HEAVY then DETERGENT is MAXIMUM
```

Table 4.1: Fuzzy control rules for a washing machine.

Weight → Dirtiness ↓	*Very_Light*	*Light*	*Heavy*	*Very_Heavy*
Almost_Clean	Very_Little	Little	Much	Much
Dirty	Little	Little	Much	Very_Much
Soiled	Much	Much	Very_Much	Maximum
Filthy	Very_Much	Much	Very_much	Maximum

The definition of the input sets, output sets, and the control rules reflect a particular point of view. If the analysis results shown in Figures 4.10 and 4.11 are not acceptable then the rules, or even the definition of the fuzzy subsets could be modified. More rules could be used for smoother operation, as well as more complex rules that include operations other than AND. The design could be expanded to include more inputs.

The diagrams for this example were generated using Fuzzy Logic Designer software developed by Byte Dynamics, Inc. Detailed discussion of software tools is presented in Chapter 9.

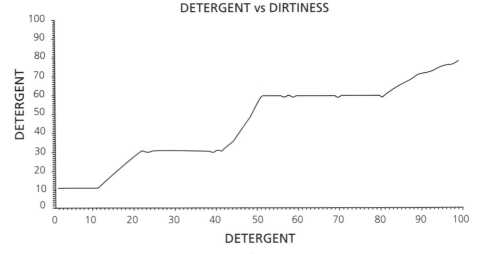

Figure 4.11: Detergent vs. Dirtiness.

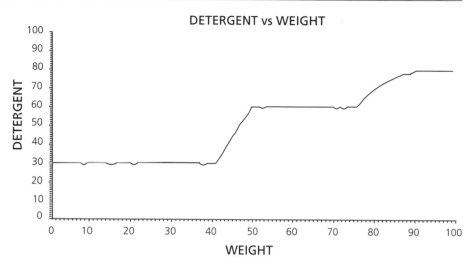

Figure 4.12: Detergent vs. Weight.

If different rules were used, different results would be obtained. For example, if the rules used are:

```
if DIRTINESS is ALMOST_CLEAN or WEIGHT is VERY_LIGHT then DETERGENT is VERY_LITTLE
if DIRTINESS is FILTHY or WEIGHT is VERY_HEAVY then DETERGENT is VERY_MUCH
if DIRTINESS is DIRTY or WEIGHT is HEAVY then DETERGENT is MUCH
if DIRTINESS is FILTHY and WEIGHT is VERY_HEAVY then DETERGENT is MAXIMUM
```

The results of the simulations would be as shown in Figures 4.13 and 4.14. One may find them objectionable (do you agree?).

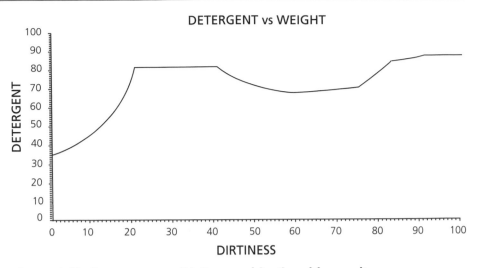

Figure 4.13: Detergent vs. Dirtiness, objectionable results.

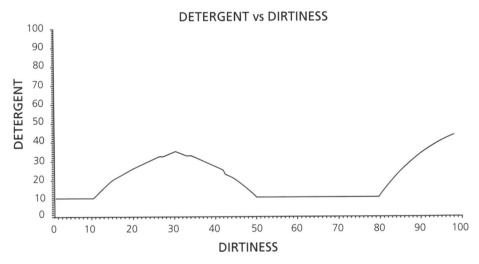

Figure 4.14: Detergent vs. Weight, objectionable results.

4.4.2 Vacuum cleaner

Once again, there are several vacuum cleaners on the market that use fuzzy logic. Links to a few of them are provided in the Web resources section at the end of this chapter. The objective of the discussion here is to highlight that the initial design steps follow the same line of reasoning as that of the washing machine. All fuzzy control applications follow a similar approach. First you need to identify the inputs and the outputs, then relate them through IF/THEN rules. The inputs here could be dirtiness, type of surface, and type of dirt, and the output could be the force of vacuuming.

One can define fuzzy subsets for the *Surface* such as : *Wood, Curtain,* and *Carpet*, and for *Dirtiness* as: *Almost_Clean, Dirty, Soiled,* and *Filthy.* The strength of vacuuming could be defined by the fuzzy subsets: *Very_Weak, Weak, Normal, Strong,* and *Very_Strong*. Rules relating the inputs and the output could be as summarized in Table 4.2. Simulation analysis could be carried out on the rules, and set definitions could be modified to make it satisfactory if it is not.

Table 4.2: Fuzzy control rules for a vacuum cleaner.

Dirtiness → Surface ↓	*Almost_Clean*	*Dirty*	*Soiled*	*Filthy*
Wood	Very_Week	Week	Normal	Strong
Curtain	Very_Week	Normal	Strong	Very_Strong
Carpet	Week	Normal	Strong	Very_Strong

Concluding Remarks

The application of fuzzy logic principles enables solving problems that would have been difficult or even impossible to solve if mathematical models were used. Fuzzy logic, however, would not create a solution if one did not exist in the first place through the knowledge of an expert; such knowledge is the basis of a fuzzy algorithm. To learn how to solve a problem is the domain of neural networks, discussed in Chapter 6. It is also important to realize that what is reached through fuzzy algorithms is *a solution*, but not necessarily the optimum one. An optimum solution could be reached from a pool of solutions using *genetic algorithms*. This topic is introduced in the Appendix.

Bibliography

1. C. von Altrock, "Fuzzy Logic in Automotive Engineering," *Circuit Cellar*, 88, 1–9, 1997.

2. J. C. Bezdek and S. K. Pal, *Fuzzy Models for Pattern Recognition*, IEEE Press, New York, 1992.

3. O. Bishop, *Understand Electronic Control Systems*, Newnes, Oxford, 2000.

4. M. M. Bourke and D. G. Fisher, "Convergence, Eigen Fuzzy Sets and Stability Analysis of Relational Matrices," Fuzzy Sets and Systems, 81, 227–234, 1996.

5. F. Bouslama and A. Ichikawa, "Application of Limit Fuzzy Controllers to Stability Analysis," *Fuzzy Sets and Systems*, 49, 103–120, 1992.

6. Y. Y. Chen, "Stability Analysis for Fuzzy Control - A Lyapunov Approach," *Proc. of IEEE Int. Conf. on Systems, Man and Cybernetics*, VA, USA, 1987, pp. 1087–1031.

7. D. P. Filev and R. R. Yager, "On the Analysis of Fuzzy Logic Controllers," *Fuzzy Sets and Systems*, 68, 39–66, 1994.

8. J. Gang and C. Laijiu, "Linguistic Stability Analysis of Fuzzy Closed Loop Control Systems," *Fuzzy Sets and Systems*, 82, 27–34, 1996.

9. O. Georgieva, "Stability of Quasilinear Fuzzy System," *Fuzzy Sets and Systems*, 73, 249–258, 1995.

10. M. M. Gupta, G. M. Trojan, and J. B. Kiszaka, "Controllability of Fuzzy Control Systems," IEEE Transactions on Systems, Man, and Cybernetics, SMC-16, 576–582, 1986.

11. L. P. Holmbald and J. J. Østergraad, "Control of a Cement Kiln by Fuzzy Logic," in *Fuzzy Information and Decision Processes*, M .M. Gupta and R. Sanchez (eds.), North Holland, 1982.

12. G. Hwang and S. Lin, "A Stability Approach to Fuzzy Control Design for Nonlinear Systems," *Fuzzy Sets and systems*, 48, 279–287, 1992.

13. C. Jianqin and C. Laijiu, " Study on Stability of Fuzzy Closed-loop Control Systems," *Fuzzy Sets and Systems*, 57, 159–168, 1993.

14. T. A. Johansen, "Fuzzy Model Based Control: Stability, Robustness, and Performance Issues," *IEEE Transactions on Fuzzy Systems*, 2, 3, 221–234, 1994.

15. A. A. Kania, "On Stability of Formal Fuzziness Systems," *Information Sciences*, 22, 51–68, 1980.

16. R. Katoh, T. Yamashita, and S. Singh, "Stability Analysis of Control Systems Having PD Type of Fuzzy Controller," *Fuzzy Sets and Systems*, 74, 321–334, 1995.

17. W. J. M. Kickert and E. H. Mamdani, "Analysis of Fuzzy Logic Controller," Fuzzy Sets and Systems, 1, 29–44, 1978.

18. P. J. King and E. H. Mamdani, "The Application of Fuzzy Control Systems to Industrial Processes," *Automatica,* 13, 232–242, 1977.

19. J. B. Kiszaka, M. M. Gupta, and P. N. Nikiforuk, "Energetistic Stability of Fuzzy Dynamic Systems," IEEE Trans. on Systems, Man, and Cybernetics, SMC-14, 6, 783–791, 1985.

20. S. Kitamura and T. Kurozumi, "Extended Circle Criterion and Stability Analysis of Fuzzy Control Systems," *Fuzzy Engineering toward Human Friendly Systems*, T. Terano, M. Sugeno, M. Mukaidono (eds.), IOS Press, 1992, pp. 634–643.

21. B. Kosko, *Fuzzy Engineering*, Prentice Hall, Upper Saddle River, New Jersey, 1997.

22. H. A. Malki, H. Li, and G. Chen, "New Design and Stability Analysis of Fuzzy Proportional-Derivative Control Systems," *IEEE Transactions on Fuzzy Systems*, 2, 4, 245–354, 1994.

23. E. H. Mamdani, "Application of Fuzzy Algorithms for the Control of a Dynamic Plant," *Proc. IEEE*, 12, 1585–1588, 1974.

24. R. J. Marks II, editor, *Fuzzy Logic Technology and Applications*, IEEE Press, New York, 1994.

25. M. Mizumoto, "Realization of PID Controls by Fuzzy Control Methods," *Proc. of the IEEE Int. Conf. on Fuzzy Systems*, San Diego, March 8–12, 1992, pp. 709–715.

26. K. S. Ray and D. D. Majumder, "Application of Circle Criteria for Stability Analysis of Linear SISO and MIMO Systems Associated with Fuzzy Logic Controllers," *IEEE Trans. on Systems, Man, and Cybernetics*, SMC-14, 2, 345–349, 1984.

27. K. S. Ray, A. H. Ghosh, and D. D. Majumder, "L_2-Stability and Related Design Concepts for SISO Linear System Associated with Fuzzy Systems," *IEEE Transactions on Systems, Man, and Cybernetics*, SMC-14, 6, 932–939, 1984.

28. L. Rezink, *Fuzzy Controllers*, Newnes, Oxford, 1997.

29. T. J. Ross, *Fuzzy Logic with Engineering Applications*, McGraw-Hill, Inc., New York, 1995.

30. K. Tanaka and M. Sugeno, "Stability Analysis and Design of Fuzzy Control Systems," *Fuzzy Sets and Systems*, 45, 135–156, 1992.

31. G. Vachtesvanos and S. Farinwata, "Fuzzy Logic Control: A Systematic Design and Performance Assessment Methodology," in *Fuzzy Logic: Implementation and Applications*, M. J. Patyra and D. M. Mlynek (eds.), John Wiley & Sons, 1996, pp. 39–62.

32. J. Wilkie, M. Johnson, and R. Katebi, *Control Engineering*, Palgrave, New York, 2002.

33. R. R. Yager and D. F. Filv, *Essentials of Fuzzy Modeling and Control*, John Wiley & Sons, Inc., New York, 1994.

34. Y. Yan, M. Tyan, and J. Power, *Using Fuzzy Logic*, Prentice Hall, New York, USA, 1994.

35. S. Y. Yi and M. J. Chung, "Systematic Design and Stability Analysis of a Fuzzy Logic Controller," *Fuzzy Sets and Systems*, 72, 271–298, 1995.

Web Resources

1. A Demonstration of Real-time Control of an Inverted Pendulum
   www.aptronix.com/fuzzynet/java/pend/pendjava.htm

   This is a demonstration of a one-stage inverted pendulum with a fuzzy Java controller provided by FuzzyNet OnLine.

2. Simulation of the Problem Fuzzy Control of a Pendulum
   www.erudit.de/erudit/demos/cartball/

   This Web site is maintained by the European Network for Fuzzy Logic and Uncertainty Modeling in Information Technology, ERUDIT. It provides a demo to illustrate the fuzzy control of an inverted pendulum. The demo can run through the Web or can be downloaded as a standalone version.

3. Fuzzy Logic On-Line Demos
   www.emsl.pnl.gov:2080/proj/neuron/fuzzy/demos.html

   This Web site provides links to numerous fuzzy demos including:

   - The Fuzzy Truck Simulator (Auburn University)

   - Fuzzy Logic Overview (Battelle Memorial Institute)

   - Amplitude Control Broom Balancing (Cambridge University)

   - Demo of Fuzzy Furnace Controller (Rick Hillegas), and more.

   The site is provided by Pacific Northwest National Laboratory, USA.

4. Fuzzy Truck Parking
   www.iit.nrc.ca/IR_public/fuzzy/FuzzyTruck.html

   The demo shows an application of fuzzy logic to park a truck by backing it up to a loading platform. It was created by the Integrated Reasoning Group of the Institute for Information Technology of the National Research Council of Canada.

5. Fuzzy Truck Simulator
   www.vgt.bme.hu/okt/tananyag/fuzzy/fuzzy12/home.html

   The simulation goal is to design a fuzzy system to back up to a truck so that it arrives at a loading dock at a right angle (relative to the horizontal axis). The applet shows that through direct manipulation of the fuzzy set rules, the system is dynamic enough to compensate to a certain degree. The source code is also provided. The site is maintained by Christopher Britton, Budapest University of Technology and Economics.

6. Tuning of Fuzzy PID Controllers
    www.iau.dtu.dk/~jj/pubs/fpid.pdf

    This is a PDF file of a 22-page paper with 10 references with the above title by Jan Jantzen, Technical University of Denmark.

7. About robustness of Fuzzy PD and PID
    www.erudit.de/erudit/events/esit2000/proceedings/BB-02-2-P.pdf

    A paper by B. Butkiewicz, Warsaw University of Technology. The main idea of the paper is to show that the behavior of a fuzzy controller depends mainly on the rules and plant parameters, not on the reasoning and defuzzification methods. The paper appeared in the Proceedings of the European Symposium on Intelligent Techniques, 2000, Aachen, Germany, September 14–15, 2000, pp. 350–356.

8. Online Papers and Reports
    www.angelfire.com/me/pcominos/paperlink.htm

    About a hundred papers and reports related to control systems available in zipped PS format or PDF format. Among them some are related to fuzzy PID control including:

    > "Fuzzy-PID Controllers vs. Fuzzy-PI Controllers" by Matilde Santos, J. M. de la Cruz, S. Dormido, and "Between Fuzzy-PID and PID-Conventional Controllers: a Good Choice" by Matilde Santos, J. M. de la Cruz, S. Dormido, A. P. de Madrid

9. Fuzzy Systems for Control Applications
    www.site.uottawa.ca/~petriu/Fuzzy-tutor.PDF

    A 30-page tutorial available in PDF format. It is authored by E. M. Petriu, University of Ottawa. The tutorial is easy to follow, very well illustrated, and provides bibliographical references.

10. Fuzzy Control Methods and their Real System Applications
    www.ic-tech.com/Fuzzy%20Logic/

    A report consisting of 67 slides presented in HTML format by IC Tech, Inc., April 2000. Topics include:

    - Fuzzy Control Methods and their Real System Applications
    - Data Acquisition and Knowledge Acquisition
    - Learning in Neural Networks
    - Boolean Logic and Fuzzy Logic
    - Fuzzy Systems
    - Design of a Fuzzy Controller
    - Automotive Applications of Fuzzy Control

11. Stability Analysis and Controller Design for Dynamic Fuzzy Systems Based on New Fuzzy Inference Approach.
www.ubka.uni-karlsruhe.de/vvv/1997/elektrotechnik/3/3.pdf

    A paper from the Elektronische Volltextarchiv, EVA (the archives of scientific publications written by members of the University of Karlsruhe). It is authored by E. Schäfers and V. Krebs. It presents a new inference method with interpolating rules as a basis for the analysis of dynamic fuzzy systems.

12. Fuzzy Logic in Embedded Microcomputers and Control Systems
www.bytecraft.com/fuzlogic.pdf

    A 48-page report by W. Banks and G. Hayward, published by Byte Craft Limited, Waterloo, Ontario, Canada and reprinted in April, 2001. The report gives a short tutorial on fuzzy logic, then discusses fuzzy logic implementation on embedded microcomputers. It also provides bibliographical references.

13. Application of Fuzzy Logic to Multimedia Technology
www.sys.uea.ac.uk/~king/restricted/

    A report in HTML format authored by M. Razaz, J. R. King, A. Duke, and A. Rogers in collaboration with BT Laboratories. The paper describes the application of fuzzy logic to determine the availability of a person within a multi-user virtual environment system. The collected information is then used to determine whether a user is there and the level of busyness of such a user. Availability is defined using fuzzy logic concepts. The paper then reports on a new fuzzy software system that has been designed and implemented to be used to determine a user's availability within the described environment.

14. A Laboratory Course on Fuzzy Control
http://eewww.eng.ohio-state.edu/~passino/PapersToPost/FC-lab-course.pdf

    A paper available in PDF format authored by S. Yurkovich and K. Passino appeared in the IEEE Transactions on Education, 42,1, 15–21, 1999. The authors describe a control laboratory course offered at the Ohio State University for senior undergraduates and graduate students.

15. Fuzzy Control of a DC Motor
http://alds.stts.edu/APPNOTE/Fuzzy/2359.PDF

    A 14-page application note in PDF format provided by STMicroelectronics. It details all aspects of a fuzzy control system for a DC motor.

16. An Approach to Motor Control with Fuzzy Logic
    http://alds.stts.edu/APPNOTE/Fuzzy/2077.PDF

    Another 14-page application note in PDF format provided by
    STMicroelectronics, discussing motor control using fuzzy logic.

17. Fuzzy Systems for Control Applications: the fuzzy backer-upper
    www.mathematica-journal.com/issue/v4i1/columns/freeman/
    64-69_Jim41.mj.pdf

    A report in PDF format by J. A. Freeman, Loral Space Information Systems.
    The report appeared in The Mathematica Journal, 4, 1, 64–69, 1994. The
    report introduces the fuzzy logic tool from the automation toolbox. A fuzzy
    control system is developed that automatically backs up a truck to a specified
    point on a loading dock.

18. Fuzzy Tracking Control Design for Nonlinear Dynamic Systems
    www.ee.nthu.edu.tw/bschen/papers/c080-fuzzy.pdf

    A paper by C.-S. Tseng, B.-S. Chen, and H.-J. Uang appeared in the IEEE
    Transactions on Fuzzy Systems, 9, 3, 381–392, 2001.

19. Texas Instruments Fuzzy Logic Control
    www-s.ti.com/sc/psheets/spra028/spra028.pdf

    A 48-page report in pdf format provided by Texas Instruments, 1993.

20. Enhanced Control of an Alternating Current Motor using Fuzzy Logic and a
    TMS320 Digital Signal Processor
    http://alds.stts.edu/APPNOTE/Fuzzy/SPRA057.PDF

    A 66-page application report in PDF format, provided by Texas Instruments,
    1995.

21. Vacuum Cleaner Refernce Platform
    http://e-www.motorola.com/brdata/PDFDB/docs/AN1843.pdf

    A 40-page Motorola Semiconductor Application Note authored by K.
    Berringer, published in the year 2000.

22. LG Electronics Fuzzy Washing Machine
    www.dreamlg.com/en/lgewasher/index.jsp

    Announcements and technical data from LG Electronics.

23. Samsung Fuzzy Washing Machine
    www.samsungelectronics.com.my/washing_machine/tech_info/index.html

    Announcements and technical data from Samsung.

24. Hitachi Microwave Oven Control
www.has.hitachi.com.sg/grp3/sicd/application/appl-pdf/28-h012.pdf

Technical information about using fuzzy logic for microwave oven control.

25. Intel Fuzzy Control Anti-lock Brakes
www.intel.com/design/mcs96/designex/2351.htm

A report in HTML format authored by D. Elting, M. Fennich, R. Kowalczyk and B. Hellenthal. Topics covered include: Fuzzy Logic Overview, Fuzzy ABS, Model BUILDER, fuzzyTECH 3.0, and Intel Fuzzy ABS Features.

26. Aptronix Fuzzy Design Notes
www.aptronix.com/fuzzynet/index.htm

These design notes include the design of a washing machine, camera auto focus, air conditioner control, and car automatic transmission.

27. fuzzyTECH Application Library/Technical
www.fuzzytech.com/

This is the Web site of fuzzyTech. It has links to: News, Products, Fuzzy Application Library, Demo, and more. The Application Library includes topics such as: What is Fuzzy Logic, Technical Applications, Business and Finance Applications, and more. The Technical Applications include: Industrial Automation, Fuzzy in Appliances, Automotive Engineering, Antilock Braking System, and more.

28. A Practical Study on the Implementation of Fuzzy Logic Controllers
http://citeseer.nj.nec.com/cordon98practical.html

A 26-page technical report with 33 references authored by O. Cordón, F. Herrera and A. Peregrín, Universidad de Granada, Spain, July 1998. Topics covered include: Fuzzy Logic Controllers, Design of Fuzzy Logic Controllers, Framework of Implementing Fuzzy Logic Controllers, Software Implementing Fuzzy Logic Controllers, and Comparative Study of Approximate and Exact Methods.

29. HP Technical Reports

A Fuzzy Inference Design on Hewlett-Packard Logic Synthesis System
by R. L. Chen
www.hpl.hp.com/techreports/93/HPL-93-70.html

Cochannel Interference in Wireless LANS and Fuzzy Clustering
by P. Golding and A. E.Jones
www.hpl.hp.com/techreports/97/HPL-97-95.html

30. Fuzzy Filtering
    www.innovatia.com/software/papers/fuzzy.htm

    An article by Dan Simon, Innovatia Software. The paper discusses fuzzy logic, a more heuristic approach to filtering. The application is fuzzy filtering for phase-locked loops; the approach can be applied to most estimation problems.

31. International Journal of Fuzzy Systems, IJFS
    www.fuzzy.org.tw/English/default.htm

    This is a publication of the Chinese Fuzzy Systems Association. All articles are in PDF format and available for free downloading. The first issue appeared in September 1999.

CHAPTER 5

Critique of Fuzzy Logic

5.1 Introduction

Most introductory writings about fuzzy logic mention that it was surrounded with controversy since its inception, then typically give no details. Critical discussions of fuzzy logic are scattered over literature that may not be accessible to practicing engineers. The limitations of fuzzy logic applications are not typically discussed either. It is, however, important for engineers to recognize and understand, to some degree, the issues that surround fuzzy logic. Here, some critical views and limitations of fuzzy logic in engineering applications are outlined in a relatively easy-to-follow approach. An extensive selected bibliography that relates to the issues discussed is provided for readers who may wish to pursue any of the issues in more depth.

5.2 Objectives of Fuzzy Logic

Numerous individuals have taken extreme positions supporting or opposing the concepts put forward by Lotfi Zadeh regarding fuzzy sets and fuzzy logic. Some maintain that probability theory can handle any kind of uncertainty; others think that fuzziness is probability in disguise, or that probability is the only sensible way to deal with uncertainty. Some found that the term *fuzzy* suits the innumerate rather than the educated. Even stronger oppositions were expressed in statements such as: *Fuzzy logic is the cocaine of science* (attributed to William Kahan), and *Fuzzification is a kind of scientific permissiveness ... I cannot conceive of fuzzification as a viable alternative to scientific method* (attributed to Rudolff Kalman). On the other extreme, some became highly enthusiastic and put forward claims that surround fuzzy logic with an aura of mystique. Some even claim that the concepts of fuzzy set theory and fuzzy logic are changing the world.

> *And so these men of Hindostan*
> *Disputed loud and long,*
> *Each of his own opinion*
> *Exceeding stiff and strong,*
> *Though each was partly in the right,*
> *And all were in the wrong!*
>
> — John Godfrey Saxe (1816–1887)

Perhaps the most reasonable critique of fuzzy logic should focus on the objectives of fuzzy logic as stated by its founder, Lotfi Zadeh, and ask: have these objectives been achieved? The objectives were to make computers think like people and to enable computing with words. When asked about the future of fuzzy logic, Professor Zadeh supposedly answered: *my crystal ball is fuzzy.* No claims about mystical nature or changing the world were made, but a distinction between vagueness and probability was emphasized.

It is obvious that computers are not yet *thinking* like people. Nor are we computing with words, but we are closer to achieving these objectives than ever before. Numerous embedded systems use fuzzy logic, alone or with other techniques, and as a result, new devices and systems have thus been enabled. One can state that, from an engineering point of view, fuzzy logic is quite successful, but like any other methodology it has its limitations. It is important for engineers to understand fuzzy logic and its limitations to get the maximum out of this important design methodology and apply it when it is advantageous, or seek an alternative otherwise. It is good to remember that:

> *Engineering is the art of doing that well with one dollar, which any bungler can do with two dollars after a fashion.*

> — Arthur M. Wellington [1847-1895], the Economic Theory of the Locomotion of Railways, Wiley, New York, 1900.

5.3 What's in a Name?

> *We had a funny discussion over the term fuzzy. The company is highly respected and asked us to find another name.*

> — Jens-Jorgen Østergaard

It is amazing to observe that because the term *fuzzy* was coined for the new concepts that were introduced, it caused strong reaction, both favorable and unfavorable. It is actually fortunate that Lotfi Zadeh coined the term *fuzzy* rather than using a less exotic term such as Empirical Computer Aided Design Methodology. There would have been less opposition, but also greatly lessened enthusiasm and interest in investigating potential applications.

One should realize that the term *fuzzy* is used as a designation for a different class of sets and a different approach to reasoning. There is nothing fuzzy about fuzzy logic. It is the logic of fuzziness, not being fuzzy itself. On the contrary, a detailed expert knowledge of the system is usually needed. It could be helpful to draw a parallel with the term *imaginary*. Rene Descartes introduced the term *imaginary* to designate a new class of numbers. He argued that *although one can imagine that every n^{th} degree equation had n roots, there were no real numbers for some of these imagined roots.* Electronics engineers realize that imaginary numbers have nothing to do with mysticism or mythology. They have real applications; imaginary is just a term that designates a different class of numbers.

5.4 Fuzziness vs. Probability

Debate began with the argument that probability theory cannot capture all aspects of uncertainty—for example, the uncertainty stemming from the intrinsic vagueness associated with human natural language. Probabilitists disagree; for example, Lavriolette and Seaman explain:

> *Probability should be conceived as a measure of an individual's uncertainty about an event or object. Probability does not exist, just as pounds or kilograms do not exist. Probability is a measure of personal uncertainty, which does exist. We contend that the adherents of FST [Fuzzy Set Theory] have failed to demonstrate that their approach solves problems which are insoluble probabilistically. FST has certainly been used to solve problems, but the fuzzy set theorists have failed to prove the existence of uncertain events which cannot be represented probabilistically. This failure coupled with their disdain for operational measures of efficacy, has led them to employ "if it works, it is good" as a measure of success. It has also led them to claim that "since probabilitists cannot solve this problem, it cannot be solved with probability." ... the need for expediency does not preclude the development of other, superior solutions.*

Lindley supported that view and added:

> *Probability is the only satisfactory measure of one's personal uncertainty about the world and I put forward the challenge that anything that can be done by alternatives to probability can be better done by probability.*

One should observe, however, that even if it is true that probability could be used to formulate human knowledge and experiences, the details have never been developed to the extent reached by fuzzy logic theory. One should also observe that fuzzy logic has a relatively short history compared to probability theory.

Fuzziness supporters point out that probability and fuzziness are related and may be complementary, but are two different concepts. Fuzziness measures the degree to which an event occurs, not whether it occurs or not as is the case with probability. Probability assumes that the event class is crisply defined and that the laws of noncontradiction $\left(A \cap \overline{A} = \varnothing \right)$ and the law of excluded middle $\left(A \cup \overline{A} = I \right)$ hold. Fuzziness occurs when these laws are violated.

Bertrand Russell pointed out much earlier that these two laws can be violated by introducing what became known as Russell's Paradox. An army barber was ordered to shave a man if and only if he does not shave himself. Who shaves the barber?

The most misleading similarity between fuzziness and probability is that both systems quantify uncertainty with numerical values in the interval [1,0]. However, it is important to note that membership values of various elements in a given fuzzy set do not have to add up to one. Another misleading similarity stems from the fact that the two systems algebraically manipulate sets and prepositions associatively, commutatively, and distributively. These similarities are misleading because a key distinction comes from the systems they are attempting to describe. For example, the uncertainty of tomorrow's weather could be stated by saying that the probability that it will rain tomorrow is 70%; by tomorrow this uncertainty changes to certainty. On the other hand, the membership value one may assign to a chair in the set COMFORT-ABLE would remain the same tomorrow. The way one thinks about the degree of comfort experienced by a given chair will not change.

The term *fuzzy probability* has been introduced in the fuzzy logic literature to describe two conceptually distinct hybrid fuzzy-probabilistic theories. The two theories have common features such as:

- they are based on and extend the classical probability theory.

- they use fuzzy numbers to quantify the probability of occurrence of an event.

- they follow almost the same probability calculus.

The two differ, however, in the assumptions they make about the origin of the uncertainty regarding the probability of a given event. The first assumes fuzziness to occur in probabilities due to fuzziness in the definition of the event in question. The second assumes fuzziness in probabilities as a way of modeling vagueness in subjective linguistic probability assignments; it could rather be referred to as linguistic probability.

5.5 Fuzzy Logic vs. Multiple-Valued Logic

Multiple-valued logics are non-classical logics; they do not restrict the number of truth values to only two. They can be categorized into families that comprise uniformly defined finite-valued and infinite-valued systems. They include:

- Lukasiewicz logics

- Gödel logics

- t-Norm related systems

- Dunn/Belnap's 4-valued system

- Product systems

Multiple-valued logic, MVL, was created as a separate subject by Jan Lukasiewicz in the 1920s. One of the main motivations of developing Lukasiewicz MVL was to create a logical system that is free of some of the paradoxes in the two-valued Boolean logic. The Lukasiewicz logic with [0,1] as the truth value set could be viewed as a starting point for fuzzy logic. Fuzzy logic can be viewed as a generalization of MVL. In fuzzy logic, one infers new facts along with their truth values. In MVL, one infers only those facts that have truth values of 1 (absolutely true). Fuzzy set theory is application oriented. It went beyond Lukasiewicz MLV by developing linguistic semantics for logic using *fuzzy hedges*. This led to a multitude of practical applications.

5.6 Philosophical Issues

No theory can ever be proven by experiment to be 'the correct' theory. However, a theory is 'falsified', when it is shown that its predictions do not agree with some experiment.

— K. R. Poper

The above statement seems to be widely accepted in engineering and philosophy. Theories, however, have their limitations and domain of applicability. The fact that a theory does not apply under certain conditions does not imply that it is not valid, but rather it is not *the* correct theory.

It is conceivable that there exists for any theory an experiment that can falsify it. But, if a theory can lead to a valid solution, it can be used for engineering purposes.

As far as the laws of mathematics refer to reality, they are not certain; and as far as they are certain, they do not refer to reality.

— Albert Einstein
Quoted in J R Newman, The World of Mathematics, 1956.

It is a fundamental engineering concept to recognize the domain of applicability of any given theory. Logic will not be deemed invalid just because the following experimental observations have been confirmed over and over again:

I returned, and saw under the sun, that the race is not to the swift, nor the battle to the strong, neither yet bread to the wise, nor yet riches to men of understanding, nor yet favour to men of skill; but time and chance happeneth to them all.

— Ecclesiastes, Chapter 9

Ellen Hisdal wrote a series of wonderful papers detailing various difficulties with fuzzy set theory. Some of them are mentioned briefly here. The details of the arguments are provided in the papers cited in the bibliography. The discussion included:

- The possibility-certainty difficulty.

  According to the theory of fuzzy sets, the numerical values of membership functions are equal to those of the possibilities for a given label. *The possibility π_x is defined to be numerically equal to the membership function of F when we are given the proposition "x is F", e.g. "John is young".* But when we know that the possibility of an event is equal to 1, we do not know whether this event is certain or only a possibility.

- The depression of the inclusive OR curve.

  The max operator is supposed to represent the inclusive OR, not the exclusive one. The depression in the resultant curve of Figure 5.1 should not be there.

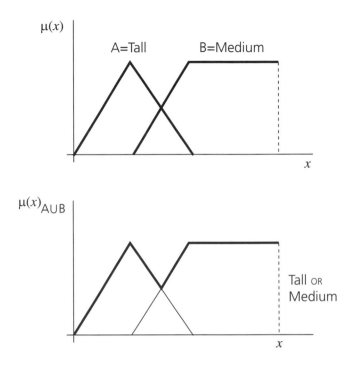

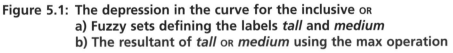

Figure 5.1: The depression in the curve for the inclusive OR
   a) Fuzzy sets defining the labels *tall* and *medium*
   b) The resultant of *tall* OR *medium* using the max operation

- The VERY operator

  The membership value of "VERY λ" is defined to be equal to the square of λ for a given value of μ. It would have been more intuitive if the definition led to an increase in the membership values resulting in an upward shift of the membership graph.

- There is a lack of distinction between the result of AND and OR at a common crossover point.

- There are too many postulates and undefined concepts.

- The shapes of the membership functions have no theoretical basis.

- The union of a fuzzy set and its complement is, in general, not equal to the attributes of the universe.

Susan Haack addressed other philosophically-oriented aspects of fuzzy logic. She criticized it directly by asking: *do we need fuzzy logic?* and reaching the conclusion that we do not. She also criticized fuzzy logic indirectly by asking: *is truth flat or bumpy?* and supported the argument that describing *p* as *one-third true* is *sheer nonsense*. She concluded that truth does not come in degrees and that fuzzy logic is not a viable competitor of classical logic.

She explained that the term *fuzzy logic* may refer to two related, but distinct, enterprises:

- The interpretation of many-valued logics in terms of fuzzy set theory. She refers to it as base logics of fuzzy logic and accepts it.

- New logical systems in which the truth values are themselves fuzzy sets. She pointed out that this does not only challenge classical logic, but also the classical conception of what logic is; she refuses that.

She suggested that it is base logics that have been given practical applications. Based on her views and arguments, we may summarize that *what is useful in fuzzy logic theory is not particularly new and what is new is not particularly useful.* We might then ask, why did the creation of so many useful applications have to wait until fuzzy logic appeared? What a coincidence! How do classical logic and reasoning explain it?

5.7 Engineering Applications Issues

Fuzzy logic, like any other methodology or theory, has its limitations and domains of applicability. The solution it provides may not be always the best solution for any problem. It is important for individuals who are interested in using fuzzy logic for practical applications to recognize that and use fuzzy logic methodology only when it is advantageous.

Limitations

Limitations of fuzzy logic from an engineering point of view include:

- Fuzzy logic cannot solve problems that have no known solution; it requires the knowledge of an expert. It is particularly useful when there is an expert who can solve the problem, but there is no mathematical model to follow. If no one knows how to solve the problem, then it follows that no rules can be devised and fuzzy logic principles cannot be applied.

■ Fuzzy logic algorithms do not have the ability to learn membership functions or rules during or after problem solving.

■ Extensive verification and validation are required, especially where safety is a key factor.

■ Assessing the stability of a fuzzy system before testing it is complicated and there is no standard, straightforward technique for stability determination.

Fuzzy control vs. PID control

Now it is the turn of engineers to make claims similar to those put forward by some mathematicians, scientists, and philosophers: *some other technique could have been used for a better result.* This time it is op-amps, resistors, and capacitors. National Semiconductor engineer and author Bob Pease explained that Mamdani's PID controller was said to give poor performance because it was misapplied. He also explained that, in general, literature that reports superior results of FLC over PID could be due to poor design of the PID.

It could have been interesting in the early days of fuzzy logic applications to compare FLC performance results with those of PID to gain confidence that FLC can work. It would still have been a useful finding if FLC was not reported to be superior to PID. But the main strength after the the technique gained some acceptance, is not in competing with PID when PID could be used satisfactorily. The aim is to design a controller where PID cannot be used because there is no simple, accurate mathematical model of the system.

Mathematical complexity

An appealing aspect of fuzzy logic from an engineering applications point of view is its stated objective of making computers think like people (not necessarily intelligent people, but those whose expertise needs to be modeled for computational purposes regardless of their intelligence level). Its approach is to model with words systems that cannot be modeled mathematically, or mathematical models that are not of practical use. It appears that fuzzy logic applications depend on common sense and modernize the old engineering methodology of design based on experience rather than theoretical analysis only: the empirical design.

Fuzzy logic, as it appears now in some literature, is quite complicated. The mathematics involved could be even more challenging than the typical mathematics needed for other engineering problem-solving methodologies. Fuzzy logic could become more complicated than it should be and more difficult to understand.

Since the mathematicians have invaded the theory of relativity, I do not understand it myself anymore.

— Albert Einstein
Quoted in P A Schilpp, Albert Einstein,
Philosopher-Scientist , Evanston ,1949.

Concluding Remarks

Fuzzy logic theory was criticized since its inception. However, its applications, particularly in embedded systems, have achieved success and increased in number over the years. This observation prompted some critiques to quip: *Fuzzy logic works in practice but not in theory!* Some over-enthusiastic supporters of the theory promote it as the answer to all problems. For engineers, it is fundamental to understand all approaches available to solve a problem, to know their merits and recognize their limitations. Only then can they make educated and practical choices regarding the optimum solution of the problem at hand. In engineering, the concern is not with a particular philosophical dogma, but with a working model. The model would be satisfactory if all observations indicate that it achieves its objectives within set limits. Practice, guided by theory, is the domain of engineering.

Bibliography

1. J. Bezdek, "Fuzziness vs. Probability – Again," *IEEE Transactions on Fuzzy Systems*, 2, 1, 1–3, 1994.

2. D. Dubois and H. Prade, "Fuzzy Sets and Probability: Misunderstandings, Bridges and Gaps," *Proc. of the 2nd IEEE Int. Conf. on Fuzzy Systems*, 2, 1059–1068, 1993.

3. C. Elkan, "The Paradoxical Success of Fuzzy Logic," *IEEE Expert*, 9, 3–8, 1994.

4. G. Epstein, *Multiple-Valued Logic Design: An Introduction*, Institute of Physics Publishing, Bristol, UK, 1993.

5. B. R. Gaines, "Fuzzy and Probability Uncertainty Logics," *Inform. & Contr.*, 38, 154–169, 1978.

6. S. Gottwald, "Many-Valued Logic," *Stanford Encyclopedia of Philosophy*, 2000 (plato.stanford.edu).

7. S. Haack, *Deviant Logic, Fuzzy Logic: Beyond the Formalism*, The University of Chicago Press, Chicago, 1996.

8. J. Halliwell and Q. Shen, "Towards a Linguistic Probability Theory," *Proc. of the 11th Int. Conf. on Fuzzy Sets and Systems, FUZZ-IEEE'02*, 2002.

9. E. Hisdal, "Infinite-valued Logic Based on Two-valued Logic and Probability. Part 1.1 Difficulties with Present-Day Fuzzy-Set Theory and Their Resolution in the TEE Model," *Int. J. man-Machine Studies*, 25, 89–111, 1986.

10. E. Hisdal, "Infinite-valued Logic Based on Two-valued Logic and Probability. Part 1.2 Different Sources of Fuzziness," *Int. J. man-Machine Studies*, 25, 89–111, 1986.

11. E. Hisdal, "The Philosophical Issues Raised by Fuzzy Set Theory," *Fuzzy Sets & Systems*, 25, 349–356, 1988.

12. E. Hisdal, "Are Grades of Membership Probabilities?" *Fuzzy Sets & Systems*, 25, 325–348, 1988.

13. G. J. Klir, 'Is There More to Uncertainty than Some Probability Theorists Might Have Us Believe?" *Int. J. General Systems*, 14, 347–378, 1989.

14. G. J. Klir, "Principles of Uncertainty: What are they? Why do we need them?' *Fuzzy Sets & Systems*, 74, 15–31, 1995.

15. B. Kosko, *Fuzzy Thinking: The New Science of Fuzzy Logic*, Hyperion, New York, 1993.

16. B. Kosko, "Fuzziness versus Probability," *Int. J. General Syst.*, 17, 211–140, 1990.

17. M. Laviolette and J. W. Seaman, Jr., "The Efficacy of Fuzzy Representations of Uncertainty," *IEEE Transactions on Fuzzy Systems*, 2, 4–15, 1994.

18. D. V. Lindley, "The Probability Approach to Treatment of Uncertainty in Artificial Intelligence and Expert Systems," *Stat Sci*, 2, 17–24, 1987.

19. D. McNeill and P. Freiberger, *Fuzzy Logic: The Discovery of a Revolutionary Computer Technology - And How it is Changing Our World*, Simon & Schuster, New York, 1993.

20. B. Pease, "What's All This Fuzzy Logic Stuff, Anyhow? *Electronic Design*, Part I, May 13, 1993, p.77, Part II, Nov. 1, 1993, p. 95, Part III, Nov. 11, 1993, p.105, Part IV, Nov. 6, 2000, Part V, Nov. 20, 2000.

21. W. Pedrycz, "Why Triangular Membership Functions?" *Fuzzy Sets & Systems*, 64, 21–30, 1994.

22. K. C. Smith, "Multiple-Valued Logic: A Tutorial and appreciation," *IEEE Computer Magazine*, 4, 17–27, 1988.

23. W. Stallings, "Fuzzy Set Theory versus Bayesian Statistics," *IEEE Trans. Syst. Man Cybern,*, 7, 216–219, 1977.

24. O. Wolkenhauer and J. M. Edmunds, "A Critique of Fuzzy Logic in Control," *Int. J. Elect. Enging. Educ.*, 34, 235–242, 1997.

25. L. A. Zadeh, "Making Computers Think Like People," *IEEE Spectrum*, 9, 26–32, 1984.

26. L. A. Zadeh, "Fuzzy Logic = Computing with Words," *IEEE Transactions on Fuzzy Systems*, 4, 2, 103–111, 1996.

27. L. A. Zadeh, "What is Fuzzy Logic and What are its Applications?" *Scientific Computing Seminar*, NERSC, May 17, 2002. (www.nersc.gov/research/SCSeminars/200220517.html).

28. L. A. Zadeh, "Some Reflections on the Anniversary of Fuzzy Sets and Systems," *Fuzzy Sets & Systems*, 100, 5–7, 1998.

CHAPTER 6

Neural Networks

6.1 Introduction

Both fuzzy logic and electronic (or artificial) neural networks are inspired by aspects of the computational power of humans. Fuzzy logic, as explained earlier, provides mechanisms for inference under cognitive uncertainty. It attempts to make computers think like people. Artificial neural networks, on the other hand, provide the means for pattern classification. They attempt to mimic the biological sensory system. The two fields developed independently, but they can complement each other in numerous applications. This chapter introduces fundamental ideas of neural networks. The following chapter complements this introduction by presenting neurofuzzy systems and their applications.

6.2 Background

Since artificial neural networks (neural networks for short) are inspired by biological systems, it is useful to give a short account of these biological systems and provide an explanation of some of the commonly used terminology. It is useful to note that the term *connectionism* appears in some relevant literature. It has its roots in an early movement in cognitive science to model the brain based on interconnection of many simple units forming a network—i.e., a neural network.

The biological neuron is the basic building unit of the nervous system. The human brain is estimated to have billions of neurons, each with thousands of connections to other neurons. It is a living cell that does not duplicate itself, yet it can develop more connections with other neurons as the individual learns more.

The main features of neurons are:

- The dendrites, which are branch-like structures that receive input signals to be processed by the cell body.
- The cell body (soma), which performs the central functions of the cell.
- The axon, which carries the output signals. It is terminated with the so-called terminal buttons. The point at which two or more neurons interact is known as the synapse. The synaptic cleft is the area between the terminal buttons of one neuron and the dendritic ends of another. The signal is transmitted across by chemicals known as neurotransmitters.

Models of biological neural networks are built around nodes (or processing units) that simulate the action of a neuron. Input and output links to these nodes simulate synapses. Weights are assigned to the input links to simulate the action of the neurotransmitters. An algorithm is then used to adjust the weights of the input links so that the neurons produce the desired output, thus simulating the process of learning. Such algorithms are therefore referred to as learning or training algorithms.

6.3 Learning

Learning, rather than being programmed, is an important feature of human computational abilities. Learning may be viewed as the change in behavior acquired due to practice or experience, and it lasts for a relatively long time. As it occurs, the effective coupling between the neurons is modified. Assessment of learning should not be limited to measuring the relatively permanent change in behavior that occurs or what the learner can do. This would be a limited view of learning. Learning may occur without any immediate measurable change in behavior. Such learning acts as a catalyst for measurable learning later. The first type of learning could be better described as training, which is usually what one hope to achieve by training an artificial neural network. The second type is the one that provides humans with the efficiency for further learning, the ability to reason, and leads to transferable skills. Both types complement each other and require time to achieve.

Learning can occur in two modes: supervised and unsupervised. In supervised learning, examples (training sets of data) are required. Unsupervised learning does not require examples; it is adaptive or self-organizing. Unsupervised learning is very common in biological systems. It is also important for artificial neural networks; training data are not always available for the intended application of the neural network.

6.4 The Basic Neuron Model

As outlined in the previous section, the biological neuron is the basic building unit of the nervous system. Similarly, the artificial neuron is the basic building unit of the artificial neural network.

It is thus important to model mathematically, and hence computationally, or using VLSI, the features of the neuron described in the previous sections. The basic features may be outlined as:

- the output from a neuron is either ON or OFF.

- the output depends on the weighted sum of the inputs. A certain threshold must be reached to make the neuron fire.

- The weights associated with the inputs model the efficiency of the synapses, coupling. A more efficient synapse will have a larger weight. The neuron is *trained* by adjusting these weights.

These features are summarized in Figure 6.1 depicting the basic neuron model.

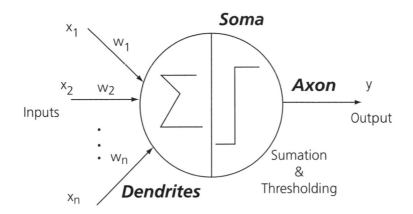

Figure 6.1: Basic neuron model. The inputs are x_1, x_2,...x_n and the output is y. The body adds the weighted inputs then compares this sum against the threshold.

From Figure 6.1 one can write:

$$\text{Total weighted input} = w_1 x_1 + w_2 x_2 + w_3 x_3 + \ldots + w_n x_n$$

$$= \sum_{i=1}^{n} w_i x_i .$$

If this sum exceeds the neuron's threshold value then the neuron fires (the output is ON; otherwise it stays OFF). This is illustrated in Figure 6.2

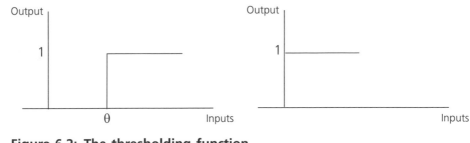

Figure 6.2: The thresholding function.
a) Thresholding at θ.
b) Thresholding at 0.

The output of the neuron can be expressed by:

$$y = f_h\left(\sum_{i=1}^{n}(w_i x_i - \theta)\right)$$

where f_h is a step function (commonly known as the Heaviside function) defined by:

$$f_h(x) = 1, x > 0$$
$$= 0, x \le 0$$

The bias, or offset , θ could be introduced as one of the inputs with a permanent weight of 1, leading to a simpler expression:

$$y = f_h\left(\sum_{i=0}^{n}(w_i x_i)\right)$$

The lower limit in the above equation is from 0 rather than 1. The value of input x_0 is always set to 1.

Using the Heaviside function leads to hard threshold. This neuron model was introduced by Warren McCulloch and Walter Pitts in the 1940's. The model is sometimes referred to as the McCulloch-Pitts neuron or MP neuron. Other functions with softer transitions are commonly used. These include the ramp and sigmoid functions. The sigmoid function can be expressed as:

$$y = \frac{1}{1 + e^{-kx}}$$

where y is the output, x is the total input and k is a gain factor that controls the sharpness of the transition from 1 to 0, as illustrated in Figure 6.3.

The sigmoid function is particularly useful because its derivative is easy to compute:

If $y = \frac{1}{1 + e^{-kx}}$, then $\frac{dy}{dx} = ky(1-y)$.

This feature could save computational time executing training algorithms.

Example

The AND Problem

Suppose the neuron is to be trained to respond to the patterns 00, 01 and 10 by an output of 0, and to the pattern 11 by an output of 1. This is the operation of a Boolean AND. The problem is illustrated in Figure 6.4. The solution is presented by the line L separating the two classes of patterns. Other lines could have been drawn to separate the patterns, i.e. there are numerous possible solutions to the problem.

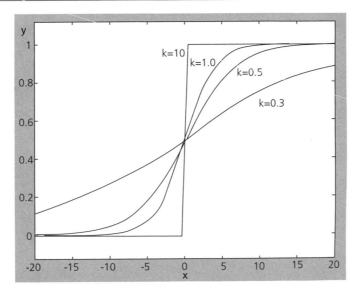

Figure 6.3: The sigmoid function.

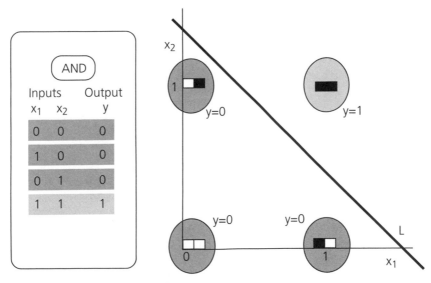

Figure 6.4: A graphical illustration of the AND problem.

A possible neuron specification to solve the problem is given in Figure 6.5. In that figure, when the input is 11 the weighted sum exceeds the threshold, leading to an output of one. It is obvious again that there are several other specifications that could have led to the same result.

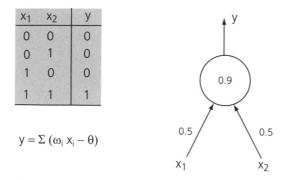

x_1	x_2	y
0	0	0
0	1	0
1	0	0
1	1	1

$$y = \Sigma \, (\omega_i \, x_i - \theta)$$

Figure 6.5: Possible specifications of a neuron to solve the AND problem. The numbers on the input arrows are weights associated with the inputs and the number inside the circle is the threshold value.

More complicated pattern recognition problems require more than one neuron. These interconnected neurons form a neural network.

6.5 The Perceptron

The perceptron is the simplest neural network. It has the structure shown in Figure 6.6. Several neurons are arranged in one layer with the inputs connected to every neuron. Learning occurs by adjusting the weights associated with the inputs so that the network can classify the input patterns. In this case a training algorithm would be necessary.

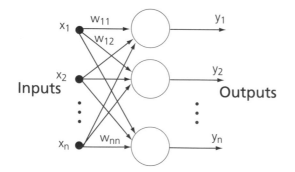

Figure 6.6: The perceptron.

The highlights of such an algorithm are:

- Initialize weights and thresholds by assigning random values.

- Present a training pattern—i.e., an input and the desired output.

- Calculate the actual output:

$$y = f\left(\sum_{i=0}^{n} w_i(t) \cdot x_i(t)\right)$$

- Adapt weights:

    If correct, then $w_i(t+1) = w_i(t)$.

    If output 0 should be 1 (class A), then $w_i(t+1) = w_i(t) + \alpha x_i(t)$.

    If output 1 should be 0 (class B), then $w_i(t+1) = w_i(t) - \alpha x_i(t)$.

    where $0 \le \alpha \le 1$ is a constant that controls the adaptation rate.

Other methods to adapt weights have been used. The Widrow-Hoff delta rule calculates the difference between the weighted sum and the desired output. This difference is called the error and given the designation, Δ. It is defined by:

$$\Delta = d(t) - y(t)$$

    where $d(t)$ is the desired response and $y(t)$ is the actual response.

Then $w_i(t+1) = w_i(t) + \alpha \Delta x_i(t)$

$d(t) = +1$ if input is from class A, and

$d(t) = 0$ if input is from class B.

Widrow called networks using this algorithm Adaptive Linear Neurons, ADALINES. Connecting many of them together leads to a structure referred to as many-ADALINES, MADALINES. The training algorithm of the perceptron attempts to determine the straight line that separates classes, but there are numerous cases where the classes are not linearly separable. The simplest example is the case of the exclusive-OR (XOR) problem. For input patterns 00 and 11, the output is to be 0 and for input patterns 01 and 10, the output is to be 1. The problem is illustrated in Figure 6.7. From that figure one can see that there is no one single line that separates the patterns. Such patterns are referred to as linearly inseparable. A single-layer perceptron cannot solve any problem where the patterns are not linearly separable.

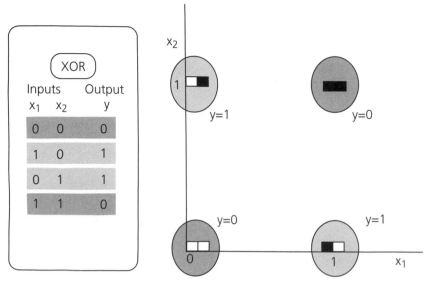

Figure 6.7: Illustration of the XOR **problem.**

6.6 The Multilayer Perceptron

To overcome the difficulty of classifying linearly inseparable patterns, a multilayer perceptron is used. A nonlinear thresholding function is also used—e.g., a sigmoid, not a step function. Layers of perceptrons with linear functions would not be more capable than a suitably chosen single layer.

A typical architecture of a multilayer perceptron is illustrated in Figure 6.8.

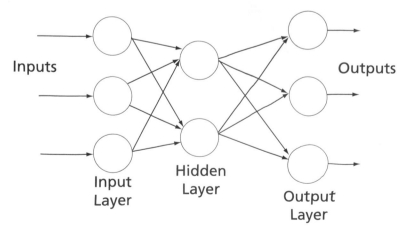

Figure 6.8: A typical architecture of a multilayer perceptron.

It is a feedforward network that consists of three layers. An input layer is directly connected to the input and an output layer is directly connected to the output. In addition, there is a layer that is not directly connected to either the input or the output and hence is referred to as the hidden layer. The hidden and output layers have a perceptron-like structure, but with the threshold function of the neurons being sigmoid rather than a step function. From this point of view, the network shown in Figure 6.8 could be referred to as a two-layer network.

The multilayer perceptron was introduced to solve the problem of linearly non-separable patterns.

Considering the XOR problem as a benchmark, it is worth investigating if a multilayer perceptron could indeed solve that problem.

Example

A possible multilayer perceptron solution to the XOR problem is shown in Figure 6.9. For input 00, the weighted sum for neuron A or neuron B is 0, the output is thus 0 and hence the weighted sum for neuron C is 0 leading to $y = 0$. For input 01, the weighted sum for neuron A is 1, and its output is 1. The weighted sum for neuron B is 1 which is less than the threshold of 1.5, thus the output is 0. The weighted sum for neuron C would be 1, leading to an output of 1. Similar reasoning applies to the input 10. For input 11, the weighted sum for neurons A and B would be 2, and hence the output of both would be 1. The weighted sum for neuron C, however, would be 0, leading to $y = 0$.

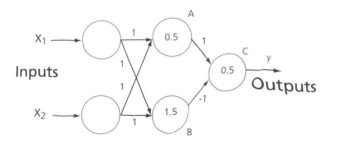

Figure 6.9: A possible solution to the XOR problem using a multilayer perceptron.

In summary, for patterns 00 and 11 the networks respond with an output of 0, and for patterns 01 and 10 the networks respond with an output of 1, thus solving the XOR problem.

One would expect that the learning rule for multilayer perceptrons would be different from that of the perceptron. The new rule is called the generalized delta rule, or the backpropagation rule. It was put forward by Rumelhart, McClelland and Williams in the 1980s. An input pattern is provided and the output is calculated and then compared with the desired output. The weights are altered accordingly so that the network produces fewer errors the next time. The use of a sigmoid function is essential so that enough information concerning the output is available to neurons in the earlier layers. This enables these neurons to get their corresponding weights adapted in order to reduce the error next time.

This type of learning is called supervised learning since a training set (input patterns and the corresponding desired output) is required. The weights of each neuron are adjusted in direct proportion to the error in the neuron to which it is connected. Backpropagating these errors through the network enables all of the weights to be correctly adapted, and thus the network learns the patterns of the training set.

The multilayer perceptron learning algorithm can be presented as follows:

- Weights and threshold are initialized by assigning to them small random values.

- A training pattern is presented consisting of an input: $x_0, x_1, x_2, \ldots, x_{n-1}$ and a target (desired) output: $t_0, t_1, t_2, \ldots, t_{m-1}$.

- The actual output of each layer is then calculated by:

$$y_{pj} = f\left(\sum_{i=0}^{n-1}(w_i x_i)\right)$$

  where y_{pj} is the output from layer j and $f(.)$ is the sigmoid function.

  The output is passed to the next layer until the final output O_{pj} is reached.

- Weights are adapted starting from the output layer and moving backwards:

$$w_{ij}(t+1) = w_{ij}(t) + \alpha \delta_{pj} O_{pj}$$

  where w_{ij} are the weights from neuron i to neuron j at time t, α is a gain factor, and δ_{pj} is the error for pattern p on neuron j.

  For the output neurons:

$$\delta_{pj} = k O_{pj}(1 - O_{pj})(t_{pj} - O_{pj})$$

For the hidden neurons:

$$\delta_{pj} = k O_{pj}(1 - O_{pj})\sum_k(\delta_{pk} w_{jk})$$

  here

     ents *summ*ation over the k neurons in the layer above neuro*n* j.

The steps are repeated until a pattern is learned. Then a new pattern is presented and so on, until all the patterns of the training set are learned.

It is useful to note that numerous variations of this algorithm have been put forward by various authors.

The training time of the network depends on the convergence rate, i.e. the speed with which the network error approaches zero. To visualize the network behavior and the possible learning difficulty one may consider the energy function defined by:

$$E = \frac{1}{2}\left(t_{pj} - O_{pj}\right)^2$$

This represents the amount by which the output deviates from the target (desired) output. Now consider a network in which only one weight is being adjusted. A possible energy function vs. weight graph is illustrated in Figure 6.10. In general, n adjustable weights lead to an n-dimensional surface. The objective of the training algorithm is to adjust the weights so that E is minimized. Each minimum (called a basin of attraction in an n-dimensional surface) represents a solution. A difficulty occurs if the network settles into a local minimum rather than a global one (B, for example, rather than A in Figure 6.10). If this occurs, the network stops learning.

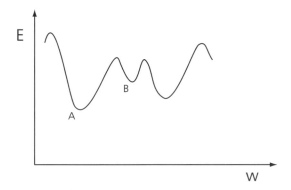

Figure 6.10: One-dimensional energy function.

A local minimum occurs when two or more disjoint classes are categorized as being the same. This difficulty could be avoided by using a training set with clearly distinguishable training examples. Adding more hidden layers reduces the occurrence of local minima. Random values may be added to the weights to make the network escape settling in a local minimum. Before using a training algorithm for the multi-layer perceptron, one has to decide the network configuration: the number of layers and the number of neurons per layer. In general, not more than two hidden layers should be needed, given enough neurons per layer. A network with the fewest possible number of neurons is an optimum network. It leads to better performance, e.g. reduction in training time, computational cost, and improved generalization capability. To reach an optimum network, there are two approaches:

- Trimming Concept

  Start with a network that has an excess of neurons and then trim it, or

- Construction Concept

  Start with a network that has a few neurons and then add more as needed.

There are several approaches to implement the trimming and construction concepts. One can examine a trained network and remove neurons with minimal contributions. One can also modify the training algorithm so that each weight is allowed to decay to zero, hence the network optimizes itself during training by removing unnecessary connections and unneeded neurons can also be removed. As well, there are algorithms that build the networks by starting with several neurons at the bottom of the architecture and then adding layers. Each added layer has fewer neurons than the previous one until the process comes to an end with a single output neuron. Other algorithms that add layers as needed have been put forward in the literature.

6.7 Recurrent Network

Learning with multilayer perceptrons discussed in the previous section is an example of supervised learning in feedforward networks—i.e., networks that allow bilateral connections between neurons. Examples of such networks are Hopfield networks and Boltzmann networks.

6.7.1 Hopfield Networks

A Hopfield network consists of a number of neurons that are fully connected. That is, every neuron is connected to every other neuron. A possible arrangement is shown in Figure 6.11. The weights are symmetrical; the weight w_{ij} connecting neuron i to neuron j is equal to w_{ji}, the weight connecting neuron j to unit i. In this model, neurons use the hard threshold function of the McCulloch-Pitts neuron. All neurons are equal in this network, but some become *more* equal after training!

The inputs to the network are applied to all neurons at the same time. The network is then given time to cycle through a succession of states until it converges to a stable solution. The stable, steady-state values of the neurons would be the output of the network.

The operation can thus be summarized as:

- All the nodes are assigned a starting value.

- An unknown input pattern is applied to all neurons.

- The network iterates to convergence.

This architecture is most suitable when exact binary (1 and −1 or 1 and 0) representations are possible such as in the case of black and white images (where input elements are pixel values), or with coded text (where input values are sets of bits). A network trained to recognize a set of patterns will be able to do so when presented with them again. Recognition will occur with the patterns even if corrupted, i.e. with the appearance not identical to that of the training set. The number of patterns such a network can memorize and remember accurately is about 15% of the number of the neurons in the network.

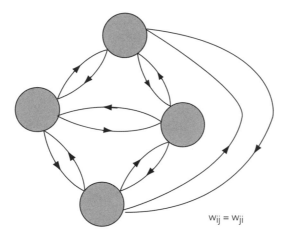

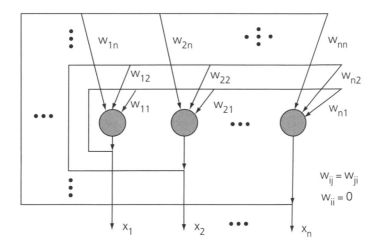

6.11: The architecture of a Hopfield network in two different arrangements.

A possible learning algorithm for the Hopfield network can be described as follows:

- Weights are assigned to all connections such that:

$$w_{ij} = \sum_{s=0}^{M-1} x_i^s x_j^s \quad \text{for } i \neq j \text{, and}$$

$$w_{ij} = 0 \quad \text{for } i = j$$

$$0 \leq i, j \leq M-1$$

where $x_i^s = \pm 1$ is element i in the training set for class s, and the training sets have M patterns.

- An unknown pattern is introduced, then

$$\mu_i(0) = x_i, \quad 0 \leq i \leq N-1$$

where $\mu_i(t)$ is the output of neuron i at time t.

- Iteration continues until steady-state is reached. The iteration is governed by,

$$\mu_i(t+1) = f\left(\sum_{i=0}^{N-1}\left(w_{ij} \cdot \mu_i(t)\right)\right), \quad 1 \leq j \leq N-1$$

- The process is repeated with other patterns in the training set.

6.7.2 Boltzmann Machine Networks

A Boltzmann machine network could be viewed as a modification of a Hopfield network. It provides a mechanism for the network to escape from local minima and move to a global minimum in the energy landscape. The idea parallels that of metallurgical annealing. In that process, a metal is heated close to its melting point then allowed to cool slowly down to room temperature. The process leads to the removal of crystal defects such as dislocations and hence allows the metal to reach a stable low-energy configuration.

In thermodynamics, the probability of a system being in state A or B follows the Boltzmann probability distribution:

$$\frac{\text{Prob}(A)}{\text{Prob}(B)} = \exp\left(-\frac{\Delta E}{T}\right)$$

where $\Delta E = E(B) - E(B)$, $E(A)$ and $E(B)$ are energies corresponding to the two states, and T is the temperature.

The Boltzmann machine network is fully connected like the Hopfield network, but unlike the Hopfield network, neurons are randomly divided into three groups. Some are chosen to be input neurons, some to be output neurons, and the rest are not connected directly to either the input or the output, i.e. they are hidden. Learning occurs in two phases: incremental and decremental.

Incremental (reinforcement)

The input and output neurons are clamped to their correct values. The network is then allowed to cycle through its states. Suppose it is the turn of the *i*th neuron, selected randomly, to evaluate its input and determine its next state. For that purpose a probabilistic rule is used. Such a method would allow transition from a higher to a lower energy state as well as a transition from a lower to a higher energy state, to allow the network to escape local minima. The probabilistic rule determines, regardless of the current state of the neuron, the next state to be 1 (or ON) with a probability, *p*, given by:

$$p = \frac{1}{\left(1 + \exp\left(-\dfrac{\Delta E}{T}\right)\right)}$$

The probability of the next state being 0 (OFF) would be, of course, $1 - p$.

T is a positive number that plays the role of temperature. It can be adjusted to have different values at various stages of converging to a steady state. ΔE is the energy of the system if the neuron is OFF and the negative value of the energy if the neuron is ON. It plays the role of activation energy. If ΔE is a positive number, then the energy of the system is higher if the neuron is OFF (at 0). Thus to minimize the energy, having the neuron ON (at 1) is preferable. Similarly, ΔE if is a negative number, then having the neuron OFF (at 0 or –1) would be the preferable state. The temperature parameter, *T*, is decremented until the output reaches a steady state. The weight between two neurons is incremented if they are both ON.

Decremental (forget bad associations)

In this phase, only the input neurons are clamped. The network is then given time to reach *thermal equilibrium* again. The weights between two neurons that are both ON are decremented. The process is repeated until the weights reach a steady-state.

6.8 Kohonen Self-Organizing Networks

6.8.1 Background

The neural networks discussed in the previous section require training patterns—that is, they have to be told the desired response for a given input and feedback obtained on their performance until they learn the patterns. Difficulty occurs if there is no training pattern or historical data for the problem the network is required to solve. A different class of networks is needed: networks that do not need examples to learn.

Such networks undergo both self-organizing and unsupervised learning. They modify the weights associated with the neural connections based on the characteristics of the input patterns. A Kohonen network is an example of such a network.

6.8.2 Description

A typical Kohonen network consists of a two-dimensional layer (or grid) of interconnected neurons (plus an input layer). All inputs are connected to every neuron and all neurons are connected to each other. The network starts with small random values assigned to the weights. All neurons receive an input pattern. Then, the winner (capitalistic neuron) takes all. Thus, neuron with the largest output and its neighbors are allowed to learn (to adapt their weights). The output of each neuron acts as an inhibitory input to other neurons but acts as an excitory in its neighborhood. This process is known as lateral inhibition. The inhibitory effect of a neuron decreases with distance from the winning neuron. This process assumes the appearance of a Mexican hat function as shown in Figure 6.12.

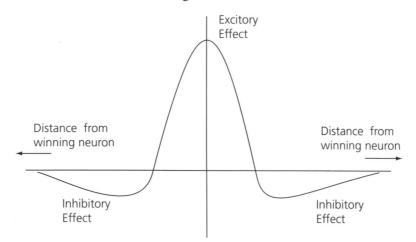

Figure 6.12: The Mexican hat function describes the relationship between the distance from the winning neuron and the connection strength.

Repetition of the process reduces the neighborhood of the winning neuron to a predetermined size. Thus, a cluster of neurons that respond to a particular pattern is formed. Repeating the process leads to converting the random set of neurons into clusters or subsets, each recognizing an input pattern.

6.8.3 Determining the winning neuron

In general, an input can be expressed as a vector with N components. It can then be normalized by dividing each of its components by the length of the vector, leading to

$$\frac{x_i}{\left(x_1^2 + x_2^2 + \ldots x_N^2\right)^{\frac{1}{2}}}$$

All normalized vectors have a distance of one unit from the origin, but in different directions. Vectors with two components will be on a unit circle. Those with three components will be on a unit sphere. Vectors with N components will be on a hypersphere of N dimensions.

Example

Suppose an input pattern A has normalized components x_1 and x_2. The input vector is compared with each of the weight vectors to find the nearest distance. The square of the distance between points (x_1,x_2) and (w_{11},w_{21}) is

$$d = \left(x_1 - w_{11}\right)^2 + \left(x_2 - w_{21}\right)^2$$
$$= \left(w_{11}^2 + w_{21}^2\right) + \left(x_1^2 + x_2^2\right) - 2\left(w_{11}x_1 + w_{21}x_2\right)$$
$$= 2 - 2\left(\text{net weighted input}\right)$$
$$= 2\left(1 - \text{net weighted input}\right)$$

The input vector and weight vector are the same when $d = 0$, i.e. when the net weighted input $= 1$. Thus, finding the nearest vector is the same as finding the neuron with the largest net input and hence the largest output.

6.8.4 Learning algorithm

The previous discussion of a Kohonen network can be summarized as follows, leading to an algorithm for self-organizing:

- weights are initialized by setting them to small random values

- a new input is presented

- the input vector $x_i(t)$ is compared with each of the weight vectors to determine which is nearest the distance d_j between the input and each output neuron j where the distance is

$$d_j = \sum_{i=0}^{n-1}\left(x_i\left(t\right) - w_{ij}\left(t\right)\right)^2$$

where $x_i(t)$ is the input neuron i at time t and $w_{ij}(t)$ is the weight from input neuron to output neuron j at time t.

- The output neuron, $j*$, with minimum distance is selected.

- The output neuron $j*$ and neighbors are updated

$$w_{ij}(t+1) = w_{ij}(t) + \eta(t)(x_i(t) - w_{ij}(t))$$

for $j \in \mathrm{NE}\, j*(t)$

where $\eta(t)$ is a gain term that decreases with time and $\mathrm{NE}\, j*(t)$ is the neighborhood of the winning neuron at time t.

- The steps are repeated by presenting a new pattern.

6.9 Adaptive Resonance Theory, ART

6.9.1 Background

Adaptive resonance theory, ART, refers to a class of self-organizing neural architectures. They include ART1, ART2, ART3, and ARTMAP. ART1 is concerned with classification of Boolean pattern sets. ART2 extends the classification to analog patterns. ART3 allows any number of ART2 modules to be concentrated into a processing hierarchy.

ART networks were introduced to solve the so-called plasticity-stability problem. It is commonly expected that people should be able to add more to what they have already learned. However, in artificial neural networks a catastrophic failure of memory may occur if we try to add more patterns to a stable, trained network. The ART networks product could be stable, yet have the ability to learn more, which is referred to as plasticity.

ARTMAP is a supervised version that can rapidly self-organize stable categorical mappings between n-dimensional inputs of m-dimensional output vectors. The key idea is to spare some output neurons for new patterns to be learned. If an input pattern and a stored pattern are sufficiently similar, they are said to be resonating and hence recognized. If they are too dissimilar, the unused output neurons are utilized to form a cluster for a new class of patterns. The network gives no response if all neurons are used.

The fundamental features of the ART network are presented in the following sections through ART1, which is historically the first (and simplest) member of the ART family.

6.9.2 ART

As opposed to the networks discussed in the previous sections, the ART network layers are not homogenous. An ART network has two layers that perform different functions. In addition, there are external parts to the layers that control the flow of the data through the network. The structure of the network is shown in Figure 6.13.

The first layer is the input/comparison layer and the second is the output/recognition layer. The neurons in the output layer are connected so that lateral inhibition can occur; that is, it uses the *winner-take-all* principle. Feedback and feedforward connections exist between every neuron in the input and output layers. Each layer has logic controls designated CONTROL-1 and CONTROL-2 connected to each neuron in layers F1 and F2, respectively.

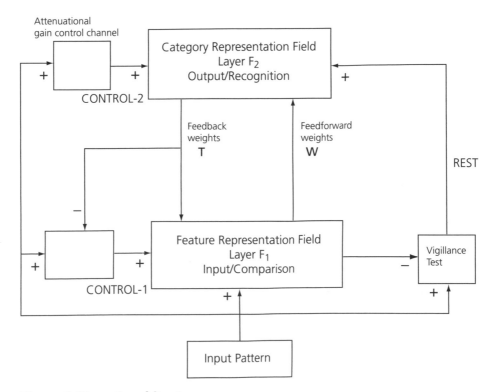

Figure 6.13: ART1 architecture.

A vigilance test evaluates the similarity of an input pattern to a stored pattern (exemplar) or template. It uses Boolean templates; each 1 in a template is referred to as a "feature". Similarities are measured by the ratio S, of the number of features common between the pattern and the template to the number of features in the input pattern, where $0 \leq S \leq 1$. For similarity to be declared: $S > \rho$, where ρ is a prescribed value referred to as the vigilance threshold. It is used to inform the network how to determine whether or not an input pattern is similar to an existing stored pattern. The value ascribed to it determines the resolution of the classification. The operation of the network can be summarized as follows:

1. **Initialization**

   - The feedback weights are set to 1, $t_{ij}(0) = 1$.
   - The feedforward weights are set such that $w_{ij}(0) = \dfrac{1}{1+N}$,

     $0 \le i \le N$, and $0 \le j \le M-1$

     where N is the number of input neurons and M is the number of output neurons.

   - The vigilance threshold, ρ, is assigned a value such that $0 < \rho < 1$.

2. **The Recognition Phase**

   - An input pattern is applied.
   - The similarity is computed:

     $$\mu_i = \sum_{i=0}^{N-1} \left(w_{ij}(t) x_i \right)$$

     where μ_j is the output of neuron j in the output layer and x_i is the input of neuron i in the input layer, which can be either 1 or 0.

   - The best matching exemplar is selected:

     $$\mu_{j*} = \max\left[\mu_j \right]$$

3. **The Comparison Phase**

   - The winning neurons feedback their pattern by adjusting the weights T_{ij} to the input.
   - The ratio S is calculated:

     $$S = \frac{\sum t_{ij} x}{\sum x_i}$$

     If $S \ge \rho$, then the classification is complete.

     If $S < \rho$, then a search phase starts.

4. **The Search Phase**

   - All outputs are reset (set to ZERO).
   - The input is reapplied.
   - The recognition and comparison phases are repeated.
   - The process is repeated until $S \ge \rho$.
   - If no classification is possible, the input is declared unknown and is allocated to a previously unassigned set of neurons in the output layer.

Concluding Remarks

In this chapter neural networks were introduced, and the concept of learning was discussed. Although there are numerous architectures and training algorithms for neural networks, it is important to observe that they all share some general features, including:

- Neural networks are trained, not programmed. Training is the process of adjusting the weights throughout the network so that it responds correctly to input patterns.

- Training may be supervised or unsupervised. Supervised training requires a training set or historical data.

- The computation depends on parallel processing. This leads to fault tolerance and graceful degradation, as opposed to catastrophic failure. The memory is not localized, but rather distributed.

- A trained network can generalize. It can classify patterns that it has not seen before; for example, a distorted input from the training set.

- The network may not always learn—i.e., the error may not approach zero.

- Time is needed for the network to learn; a network with complex functions would take longer to learn. The learning time also depends on the architecture and learning algorithm used.

- The network may over-learn, or learn the noise associated with input patterns.

- The capacity of the network is, in general, dependent on the number of neurons used.

Bibliography

1. A. Aarts and J. Korst, *Simulated Annealing and Boltzmann Machines*, John Wiley, New York, 1989.

2. I. Aleksander and H. Morton, *An Introduction to Neural Computing*, ITP Press, London, 1991.

3. D. J. Amit, *Modeling Brain Function*, Cambridge University Press, 1989.

4. R. Beale and J. Jackson, *Neural Computing*, Adam Higler, Bristol, 1991.

5. R. Bharath and J. Drosen, *Neural Network Computing*, Windcrest/McGraw-Hill, New York, 1994.

6. G. A. Carpenter and S. Grossberg, "ART-2: Self-organization of Stable Category Recognition Codes for Analog Input Patterns," *Applied Optics*, 26, 4919–30, 1987.

7. G. A. Carpenter and S. Grossberg, "The ART Adaptive Pattern Recognition by Self-organizing Neural Networks," *IEEE Computer*, 21, 77–90, 1988.

8. G. A. Carpenter and S. Grossberg, "ART3: Hierarchical Search using Chemical Transmitters in Self-organizing Pattern Recognition Architectures," *Neural Networks*, 3, 129–52, 1990.

9. G. A. Carpenter, S. Grossberg, and J. H. Reynolds, "ARTMAP: Supervised Realtime Learning and Classification of Nonstationary Data by Self-organizing Neural Network," *Neural Networks*, 4, 565–88, 1991.

10. M. Caudill and C. Butler, *Naturally Intelligent Systems*, The MIT Press, Cambridge, 1990.

11. M. Chester, *Neural Networks: A Tutorial*, PTR Prentice Hall, Englewood Cliffs, NJ, 1993.

12. S. Deutsch and A. Deutsch, *Understanding the Nervous System: An Engineering Perspective*, IEEE Press, New York, 1993.

13. K. Gurney, *An Introduction to Neural Networks*, Routledge, Taylor & Francis Group, London, 1999.

14. M. T. Hagan, H. B. Demuth, and M. Beale, *Neural Network Design*, PWS Publishing Company, Boston, 1996.

15. S. Haykin, *Neural Networks: A Comprehensive Foundation*, Macmillan College Publishing Company, New York, 1994.

16. J. Hertz, A. Krogh, and R. G. Palmer, *Introduction to the Theory of Neural Computing*, Addison-Wesley Publishing Company, New York, 1991.

17. J. J. Hopfield, "Neural Networks and Physical Systems with Emergent Collective Computational Properties," *Proceedings of the National Academy of Sciences of the USA*, 79, 2554–88, 1982.

18. T. Kohonen, "The Self-organizing Map," *Proceedings of the IEEE*, 78, 1464–80, 1990.

19. R. P. Lippmann, "An Introduction to Computing with Neural Networks," *IEEE ASSP Magazine,* 4, 22, 1987.

20. W. T. Miller III, R. S. Sutton, and P. J. Werbos (eds.), *Neural Networks for Control*, The MIT Press, Cambridge, 1995.

21. P. K. Simpson (ed.), *Neural Networks Theory, Technology, and Applications*, The IEEE Press, New York, 1996.

22. D. M. Skapura, *Building Neural Networks*, Addison-Wesley Publishing Company, New York, 1995.

23. Y. X. Zhong, "How is the Consciousness Related to Neural Networks?," *Proceedings of the 8th International Conference on Neural Information Processing,* Shanghai, China, Nov. 14–18, 2000, pp.1–4.

Web Resources

1. Artificial Neural Networks Tutorial
    www.gc.ssr.upm.es/inves/neural/ann1/anntutorial.html

    A tutorial provided in three languages by Dr. Fuente and presented in html format. It covers basic concepts of neural networks, supervised and unsupervised models. It also provides examples and a bibliography.

2. Neural Networks Tutorial with Java Applets
    diwww.epfl.ch/mantra/tutorial/english/

    An excellent tutorial of neural networks through Java applets. It is provided by Laboratoire de Microelectronique, École Polytechnique Fédérale de Lausanne. The topics include:

    - Single Neurons
      – Artificial Neuron
      – McCulloch-Pitts Neuron
      – Spiking Neuron. (Requires Swing)
      – Hodgkin-Huxley Model
      – Axons and Action Potential Propagation
    - Supervised Learning
    - Single-layer networks
    - Multi-layer networks
    - Density Estimation and Interpolation
    - Unsupervised Learning
    - Reinforcement Learning
    - Network Dynamics
      – Hopfield Network.
      – Pseudoinverse Network.
      – Network of spiking neurons. (Requires Swing).
      – Retina Simulation. (Runs very slow with some Netscape versions).

3. The Basics of Neural Networks Demystified
    www.contingencies.org/novdec01/workshop.pdf

    An essay with eight references by Louise Francis. The essay appeared in *Contingencies*, Nov/Dec, pp. 56–61, 2001.

4. Neural Networks Report
   www.doc.ic.ac.uk/~nd/surprise_96/journal/vol4/cs11/report.html

   An introductory report on Neural Networks by Christo Stergiou and Dimitrios Siganos, Department of Computing, Imperial College, London. It has seventeen references, five of which are Web accessible. The report topics include: neuron models, the architecture of neural networks, the learning process, and applications of neural networks.

5. Neural Nets
   www.shef.ac.uk/psychology/gurney/notes/index.html

   A book-size tutorial by Kevin Gurney, Department of Psychology, University of Sheffield, UK. It is available in HTML and PS formats. Topics covered include:

   Computers and Symbols versus Nets and Neurons, Learning Rules, The Delta Rule, Multilayer Nets and Backpropagation, Hopfield Network, Competition and Self-organization, and more.

6. An Introduction to Neural Networks
   www.cs.stir.ac.uk/~lss/NNIntro/InvSlides.html

   This is an introduction to neural networks in HTML format based on a talk given by Prof. Leslie Smith, Centre for Cognitive and Computational Neuro-science, Department of Computing and Mathematics, University of Stirling, UK. It provides a very useful overview of neural networks, reviews some algorithms and architectures, and discusses their past applications and likely new applications.

7. Connectionism
   plato.stanford.edu/entries/connectionism/

   An essay from the Stanford Encyclopedia of Philosophy. It defines the term and discusses related topics. They include:

   - A Description of Neural Networks,
   - Neural Network Learning and Backpropagation,
   - Samples of What Neural Networks Can Do,
   - Strengths and Weaknesses of Neural Network Models,
   - Connectionist Representation,
   - The Shape of the Controversy between Connectionists and Classicists,
   - The Systematicity Debate,
   - Connectionism and Semantic Similarity, and
   - Connectionism and the Elimination of Folk Psychology.

A bibliography and Internet resources are available.

8. Lecture on Connectionism
   yoda.cis.temple.edu:8080/UGAIWWW/lectures95/meeden/nnet/nnet.html

   Lecture note on Connectionism by Nikos Drakos of the Computer Based Learning Unit, University of Leeds. Topics covered include: an introduction to connectionism, learning to read with NETtalk, and suggestions for laboratory experiments. It also provides ten references.

9. Connectionism, Confusion, and Cognitive Science
   www.bcp.psych.ualberta.ca/~mike/Pearl_Street/ Papers/Confuse/ confuse.html

   An HTML version of a paper by Michael R.W. Dawson (Biological Computation Project, Department of Psychology, University of Alberta, Edmonton, Canada) and Kevin S. Shamanski that appeared in The Journal Of Intelligent Systems, 1994. It discusses the topic from a non-engineering point of view. It concludes that connectionism produces systems that generate interesting and sophisticated behaviors. However, connectionism has little to offer in terms of theoretical accounts of the internal structures, procedures, and representations that produce this behavior. It leads to the possibility of constructing intelligence without first understanding it.

10. Neural Networks Papers
    www.dsp.pub.ro/articles/articles.htm

    A collection of Web links to presentations on topics that include:

    - History of neural networks
    - The connectionist paradigm
    - Logic function modelling using neural networks
    - Continuous function approximation
    - Neural networks learning methods and algorithms
    - Neural networks back propagation algorithm
    - Neural networks competitive learning
    - Self organizing feature maps
    - Neural networks learning transformations. Reinforcement learning
    - Neural networks knowledge extraction
    - Networks of spiking neurons
    - Boltzmann machines
    - Belief networks
    - Radial basis functions

    The site is maintained by the DSP Laboratory, Department of Electrical Engineering, University of Politechnica, of Bucharest, Romania.

11. Artificial Neural Networks Technology
www.dacs.dtic.mil/techs/neural/neural_ToC.html

    A report on neural networks by Dave Anderson and George McNeil, Data & Analysis Center for Software, New York, USA. It is available in HTML, PDF, PS, and text formats. Topics presented include:

    - Operations
    - Training an Artificial Neural Network
    - How Neural Networks Differ from Traditional Computing and Expert Systems
    - History of Neural Networks
    - Detailed Description of Neural Network Components and How They Work
    - Network Selection:
      – Networks for Prediction
      – Networks for Classification
      – Networks for Data Association
      – Networks for Data Conceptualization
    - Networks for Data Filtering
    - How Artificial Neural Networks Are Being Used
    - Emerging Technologies

12. Neural Networks Glossary
rfhs8012.fh-regensburg.de/~saj39122/jfroehl/diplom/e-glossary.html

    A small glossary on neural network terms. Entries are accessible alphabetically, and some have hyperlinks. It is maintained by Joehen Fröhlich, Farchhochschule Regensburg, Germany.

13. Glossary of Neural Computing
www.brainstorm.co.uk/NCTT/portfolo/part06/welcome.htm

    A list of about 70 terms related to neural computing, maintained by DTI NeuroComputing Web.

14. Pictures of the Biological Neurons
glinda.lrsm.upenn.edu/~weeks/neurons.html

    A collection of pictures of biological neurons provided by Eric Weeks, Emory University, Atlanta, GA, USA.

15. Neurons Gallery
    faculty.washington.edu/chudler/gall1.html

    A collection of neurons in colour with information provided about each. The page is maintained by Eric Chudler, Department of Anesthesiology, University of Washington, USA.

16. Freeware Tools for Neural Networks
    vortex.cs.wayne.edu/NeuralNets/freelist.htm

    This site provides a list of software related to neural networks. Some are free, others are not. The list includes:

    @BRAIN, 1st Class, Adaptive Logic Network Educational Kit, AIM Estimator , AIM Problem Solver, The ART Gallery, Aspirin/Migraines, Baysian learning for neural networks, BrainMaker, CAD/Chem-Customer Formulation System, Cascade Neural Network Simulator, Cascor, Fuzzy ART, Neural Networks at your Fingertips, NeuralShell, and much more.

17. Neural Networks Software
    www.ncrg.aston.ac.uk/NN/software.html

    A big list of software related to neural networks with short descriptions. The list is provided by the Neural Computing Reseach Group, NCRG, Aston University, Birmingham, UK.

18. Neural Networks Warehouse
    neuralnetworks.ai-depot.com/Applications.html

    This site provides links under numerous categories including: books, tutorials, libraries, software, programs, research, and applications. The site is owned and maintained by Alex J. Champandard, Artificial Intelligence Depot.

19. Bibliographies on Neural Networks
    liinwww.ira.uka.de/bibliography/Neural/index.html

    A huge collection of papers related to neural networks. The collection is searchable by title. It is also divided into sub-groups including:

    - Bibliography on the Self-Organizing Map (SOM) and Learning Vector Quantization (LVQ)
    - Bibliography for the *IEEE Transactions on Neural Networks*
    - Bibliography for *Advances in Neural Information Processing Systems*
    - Bibliography for the journal: *Neural Computation*
    - Bibliography for the journal: *Neural Networks*
    - Bibliography on Neurofuzzy Systems

- Bibliography on constructive algorithms for neural networks
- Bibliography on Recurrent Neural Networks
- Bibliography on Invariant Pattern Recognition with Neural Networks
- Bibliography of Fault Tolerance related Neural Network literature
- Bibliography on Adaptive Resonance Theory (ART)
  and much more.

CHAPTER 7

Hybrid Systems

7.1 Introduction

Fuzzy logic systems and neural networks were inspired by human computational abilities. Thus, they share some common ground. For example, they do not require mathematical models and they work under imprecise, uncertain, and noisy environments. It is also interesting to observe that both are numerical in nature, although this is not a human trait. They also have some complementary characteristics. Fuzzy logic provides inference mechanisms under cognitive uncertainty; it enables computing with words. Fuzzy logic systems require rules set by a human expert; they cannot learn by example or on their own. In contrast, neural networks can learn, adapt, and generalize. In addition, neural networks exhibit fault tolerance due to their distributed structures.

Neural methods can be incorporated into fuzzy systems and fuzzy logic methods can be introduced into neural networks to take advantage of their complementary characteristics. Such combinations lead to what is referred to here as hybrid systems, which are also called fuzzy-neural, neural-fuzzy, or fuzzy-neuro systems. Such systems combine the merits of both fuzzy logic systems and neural networks; namely, the explicit knowledge representation from fuzzy logic methods and the learning ability of neural networks. Using this combination of methods, a more versatile system with a human-like approach can be designed.

7.2 Fuzzy Neuron

There are numerous possible ways to introduce fuzzy logic methods into a neural network. For example, the model of the neuron discussed in Section 6.1 could be generalized to incorporate fuzziness leading to a *fuzzy neuron*. It is also possible to introduce fuzziness in a network with non-fuzzy neurons by modifying the learning algorithms to allow for membership functions. In this section the concept of fuzzy neurons is illustrated.

A possible model for a fuzzy neuron, FN, that can express and process fuzzy information was put forward by Kwan and Cai. It is illustrated in Figure 7.1. The FN has inputs x_i associated with weights w_i ($i = 1$ to n). It also has outputs y_j ($j = 1$ to m); all of the outputs have values in the interval [0, 1]. These could represent membership values of a given pattern to a particular fuzzy set. The inputs may be thought of as the representation of a linguistic variable and the output expresses the membership values of assigned linguistic descriptions such as TALL, MEDIUM, SMALL, etc. These values could then be propagated to other neurons.

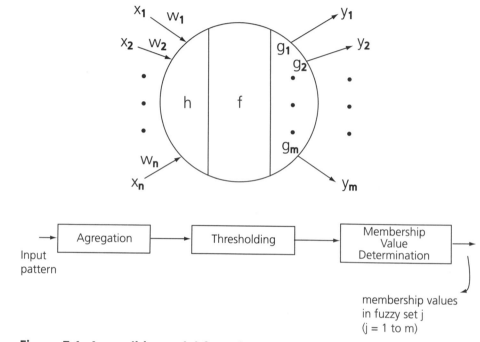

Figure 7.1: A possible model for a fuzzy neuron.

In mathematical terms, the FN operation could be described by

$$y_j = g_i \left[f \left(h_{i=1}^n (w_i - x_i) - \theta \right) \right]$$

where h is an aggregation function (such as MIN, MAX, etc.) that replaces the summation operation in a non-fuzzy neuron model, f is an activation function, θ is the activation threshold, and $\{g_i, j = 1, 2, \ldots, m\}$ are m output functions that represent the membership functions of the input pattern in all the m fuzzy sets.

If z is the aggregated weighted input $h_{i=1}^{n}\left(w_i x_i\right)$, then one could define:

- INPUT-FN with which an input layer is composed such that $z = x$.

- MAX-FN in which the aggregation function is a MAX operation, leading to

$$z = \max_{i=1}^{n}\left(w_i x_i\right)$$

  Such a neuron could be referred to also as an OR-FN.

- MIN-FN in which the aggregation function is a MIN function, leading to

$$z = \min_{i=1}^{n}\left(w_i x_i\right)$$

  Such a neuron could be referred to as an AND-FN.

- COMP-FN (competitive FN) in which the activation threshold, θ, is a variable and there is only one output such that

$$y = g\left(s - \theta\right) = 0, s < \theta$$
$$= 1, s \geq \theta$$

  $s = f\left(z - \theta\right)$ defines the state of the FN, and

  $\theta = t\left(c_1, c_2, \ldots, c_K\right)$, t being the threshold function, and

  $c_k\left(k = 1 \text{ to } K\right)$ are competitive variables of the FN.

All these neurons could appear in one fuzzy neural network, FNN, leading to a heterogeneously structured network as opposed to the homogeneous structures, where all the neurons have similar definitions, as discussed in Chapter 6.

7.3 Multilayer FNN Architectures

A multilayer computational structure consisting of fuzzy neurons can be described as a multilayer fuzzy neural network. Numerous architectures have been suggested in the literature. The following two examples are provided to illustrate the concept.

Example

Pedrycz Structures

A simplified three-layer fuzzy neural network structure is shown in Figure 7.2. Each layer consists of the same type of FNs, but different types are used for each layer. The input layer consists of INPUT-FNs, the hidden layer of MAX-FNs, and the output layer of a single MIN-FN. Figure 7.3 illustrates an alternative structure where the hidden layer is composed of MIN-FNs, and the output layer is a single MAX-FN.

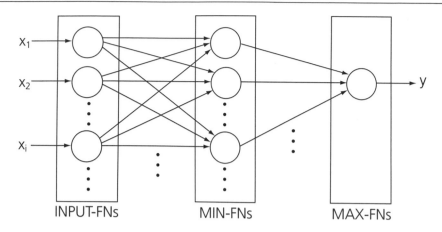

Figure 7.2: A simplified three-layer fuzzy neural network with a hidden layer composed of MIN-FNs.

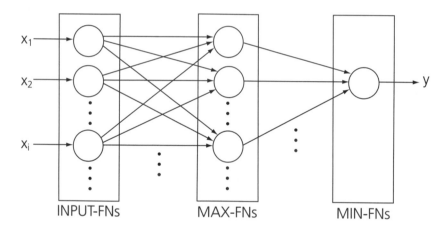

Figure 7.3: A simplified three-layer fuzzy neural network with a hidden layer composed of MAX-FNs.

Example

Kwan-Cai Structure

A four-layer feedforward fuzzy neural network, FNN, for 2-D pattern processing is shown in Figure 7.4. The first layer of the network accepts pixel values of an input pattern and transfers them into normalized values in the interval [0, 1]. The second layer fuzzifies the input pattern to an adjustable predetermined degree such that all distinct training patterns can be separated by the network and it produces an acceptable recognition rates. The mth output of this layer can be thought of as expressing a fuzzy concept. The third layer gives similarity of the input pattern to all learned patterns. The fourth layer is a defuzzier stage. It chooses the maximum similarity as the activation threshold of all the COMP-FNs.

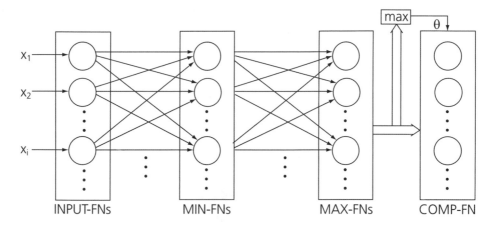

Figure 7.4: Kwan-Cai FNN structure.

7.4 Fuzzy ART

Adaptive Resonance Theory neural networks, ART, can be built on prior learning as outlined in Chapter 6. They can create new categories to accommodate inputs that do not belong to any of the categories learned before. Fuzzy ART generalizes ART1 to incorporate fuzzy computations. The generalization to learning both analog and binary input patterns is achieved by replacing the classical intersection operator in ART1 calculations with the fuzzy AND operation—that is, the MIN operation. Fuzzy ART still clusters patterns (vectors) by comparing an input pattern to existing ones, then creating a new category if the need arises. The highlights of the model put forward by Carpenter et al. are as follows:

- Three parameters that control the dynamics of the fuzzy ART are defined: the choice parameter, $\alpha > 0$, the learning rate parameter, $\beta \in [0,1]$, and the vigilance parameter, $\rho \in [0,1]$.

- For each M-dimensional input $\mathbf{I} = (I_1, I_2, \ldots, I_M)$ and cluster category j, a choice function is defined by

$$T_j(\mathbf{I}) = \frac{|\mathbf{I} \wedge \mathbf{w}_j|}{\alpha + |\mathbf{w}_j|}$$

where $\mathbf{w}_j = (w_{j1}, w_{j2}, \ldots, w_{jM})$ is the weight vector associated with category j. The number of potential categories N is arbitrary. The fuzzy ART weight vector $\mathbf{w}_j$ represents the bottom-up and top-down weight vectors of ART1, the operator $\wedge$ denotes a fuzzy AND (i.e., MIN) operation, and the norm $|\ldots|$ is defined by:

$$|x| = \sum_{i=1}^{M} |x_i|.$$

$T_j(\mathbf{I})$ is simply written as T_j when $\mathbf{I}$ is fixed.

- The category choice is indexed by J, where:

$$T_J = \max\left[T_j, j = 1, 2, \ldots, N\right]$$

If more than one T_j is maximal, the category j with the smallest index J is chosen so that neurons become committed in order $j = 1, 2, 3, \ldots$.

- The matching function is defined by:

$$S_j = \frac{|\mathbf{I} \wedge \mathbf{w}_j|}{|\mathbf{I}|}$$

If $S_j \geq \rho$, resonance occurs and the learning process starts.

If $S_j < \rho$, mismatch reset occurs and T_j is reset to -1 for the duration of the input presentation. A new index J is chosen, and the search continues until resonance occurs.

- Learning is achieved by updating the weight vector, $\mathbf{w}_J$.

$$\mathbf{w}_J^{\text{new}} = \beta\left(I \wedge \mathbf{w}_J^{\text{old}}\right) + (1 - \beta)\mathbf{w}_J^{\text{old}}.$$

- It was reported to be advantageous to normalize the inputs. This could be achieved by processing each input vector $\mathbf{a}$ by setting $\mathbf{I} = \dfrac{\mathbf{a}}{|\mathbf{a}|}$. An alternative normalization rule known as *complement coding* could also be used. If a and a^c represent the ON the OFF responses, respectively, then $a_i^c \equiv 1 - a_i$. The complement coded input $\mathbf{I}$ becomes a 2M-dimensional vector $\mathbf{I} = (a, a^c)$. Since

$$\mathbf{I} = \left(a, a^c \right)$$

$$= \sum_{i=1}^{M} a_i + \left(M - \sum_{i=1}^{M} a_i \right)$$

$$= M$$

then inputs that are complement coded are normalized.

7.6 Fuzzy ARTMAP

A fuzzy ARTMAP, as described by Carpenter et al., is a system that can learn to classify inputs by a fuzzy set of features, or a pattern of fuzzy membership values expressing the extent to which each feature is present.

An illustration of the system is shown in Figure 7.5. The system consists of two fuzzy ART modules linked by an inter-ART module, called a *map field*, F^{ab}. It is used to form predictive associations between categories and carry out the *match tracking rule*: the vigilance parameter of ART-a increases in response to predictive mismatch at ART-b. Match tracking recognizes category structure so that predictive errors are not repeated when the input is presented again.

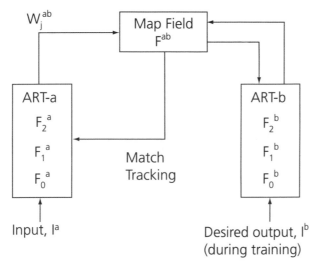

Figure 7.5: Fuzzy ARTMAP architecture.

The operational highlights of the network are as follows:

- Fields F_0^a and F_0^b complement-code the M_a-dimensional vector **a** and the M_b-dimensional vector **b**.

  For ART-a: $\mathbf{I}^a = \mathbf{A} = (a, a^c)$, and for ART-b: $\mathbf{I}^b = \mathbf{B} = (b, b^c)$.

- For ART-a, the output of the field F_1^a is denoted by $\mathbf{x}^a \equiv \left(x_1^a, x_2^a, \ldots, x_{2Ma}^a \right)$, the output of field F_2^a is denoted by $\mathbf{y}^a \equiv \left(y_1^a, y_2^a, \ldots, y_{2Ma}^a \right)$, and the jth ART-a weight vectors are denoted by $\mathbf{w}_j^a \equiv \left(w_{j1}^a, w_{j2}^a, \ldots, w_{2Ma}^a \right)$.

- For ART-b, $\mathbf{x}^b \equiv \left(x_1^b, x_2^b, \ldots, x_{2Mb}^b \right)$ denote the output vector of field F_1^b, and $\mathbf{y}^b \equiv \left(y_1^b, y_2^b, \ldots, y_{Nb}^b \right)$, and $\mathbf{w}_k^b \equiv \left(w_{k1}^b, w_{k2}^b, \ldots, w_{2Mb}^b \right)$.

- For the map field, $\mathbf{x}^{ab} \equiv \left(x_1^{ab}, x_2^{ab}, \ldots, x_{Nb}^{ab} \right)$ denote the F_{ab} output vector, $\mathbf{w}_j^{ab} \equiv \left(w_{j1}^{ab}, w_{j2}^{ab}, \ldots, w_{jNb}^{ab} \right)$ denotes the weight vector from the jth F_2^a neuron to F^{ab}.

- The map field is activated whenever one of the ART-a or ART-b categories is active. If both ART-a and ART-b are active, then the field map becomes active only if ART-a predicts the same category as ART-b.

- At the start of each input presentation, the ART-a vigilance parameter ρ_a is set to a baseline value. If $\left| \mathbf{x}^{ab} \right| < \rho_{ab} \left| \mathbf{y}^b \right|$, where ρ_{ab} is the map field vigilance parameter, then ρ_a is increased until $\rho_a |A| > |A \wedge w_j^a|$, where A is the input in complement form to F_2^a and J is the index of the active F_2^a neuron. When this occurs, the ART-a search leads to either the activation of another F_2^a node or the shutdown of F_2^a for the rest of the input presentation.

- Learning rules determine how the map field weights w_{jk}^{ab} change.

7.7 Neural Fuzzy Systems

In fuzzy set theory, the universe of discourse needs to be divided into fuzzy subsets. Each subset is defined by a membership function. This process requires the knowledge of a human expert. On the other hand, neural networks perform conceptually similar tasks. They can divide the data space into clusters based on certain features. In a fuzzy system context, clustering multidimensional data could lead to:

- One-dimensional membership functions based on a given metric, or

- Multidimensional membership functions that model fuzzy relations—in other words, fuzzy IF/THEN rules.

Thus, a neural network could identify and extract membership functions and fuzzy rules without the knowledge of a human expert. One may ask: how this could be useful?

Recall that a basic idea of fuzzy logic is to map human thinking into a computer algorithm. It is possible that the physical environment under consideration may change over a period of time. Adapting of the fuzzy system through adjusting membership functions and fuzzy rules would be needed to better reflect the changes occurring in the actual system.

Neural networks could also perform computationally intensive tasks, such as fuzzy logic inference. Knowledge of the rules (the antecedent and consequent clauses) is encoded in the network. Keller et al. suggested a four-layer network to do that task. The first is an input layer, the second is for checking the antecedent clause, the third is for clause combining, and the fourth layer is the output layer.

Various methods and algorithms have been suggested in the literature. The following example illustrates an algorithm put forward by Adeli and Hung for determining the membership function.

Example

Membership determination using the Adeli-Hung algorithm (AHA).

The neural network for this algorithm has a flat topology—i.e., two layers including the input layer. The number of input neurons equals the number of patterns, M, in each training instance. There are N training instances: $X_1, X_2, \ldots, X_N$. The number of output neurons equals the number of clusters, N.

The algorithm starts with M inputs and one output. The network can be described in shorthand by NN(M,1); this reads as *a neural network with m inputs and one output*. The final number of output neurons is yet to be determined. The first training instance gets assigned to the first cluster. Then, for the second instance:

- If it is classified to the first cluster, the output neuron representing the first cluster becomes active.

- If it is classified as a new cluster, an additional output neuron is added to the network.

The process continues until all training instances are classified.

The details of the AHA can be highlighted as follows:

- The mean vector of the training instance is defined by:

$$\bar{X}_N = \frac{1}{N} \sum_{i=1}^{N} x_i .$$ It follows that for the (N + 1) training instance:

$$\bar{X}_{N+1} = \frac{N}{N+1} \bar{X}_N + \frac{N}{N+1} X_{N+1} .$$

- A function giving the degree of difference is defined using Euclidean distance as:

$$diff\left(X,C_i\right)=\sqrt{\sum_{j=1}^{M}\left(x_j-c_{ij}\right)^2}\ .$$

  It represents the difference between a training instance, X, and a cluster C (mean of the patterns of the instances in the cluster) in an NN(M,P) network (a neural network with M input neurons and P output neurons).

- The cluster with the smallest degree of difference $diff_{min}\left(X,C_i\right)$ is made active.

$$C_{active}=\left\{C\big|\min\left[diff\left(X,C_i\right),i=1,2,...,p\right]\right\}.$$

- If the clusters are completely disjoint, a binary matrix Z is used to record the cluster of each instance:

  $z_{ij}=1$, if instance i belongs to cluster j, and
  $z_{ij}=0$, if instance i does not belong to cluster j.

- If the clusters are partly overlapping, the boundaries of the clusters become fuzzy. A binary matrix is still used. The degree of membership of each instance in the cluster is based on the distance between the instance and the prototype of the cluster. The prototype for each cluster is defined as the mean of all instances in that cluster.

- Assuming n_p instances in a cluster p, the pattern vector of the instance number i in that cluster is

$$X_i^P=\left(x_{i1}^P,x_{i2}^P,...,x_{iM}^P\right),\text{ and}$$

$$C_p=\left(c_{p1},c_{p2},...c_{pM}\right)$$

$$=\frac{1}{n_p}\sum_{i=1}^{n_p}X_i^P.$$

  where $c_{pi}=\dfrac{1}{n_p}\sum_{i=1}^{n_p}x_{ij}^P$, and $j=1,2,...,M$.

- Assuming triangular membership functions, the membership value of the instance number i in the p cluster is defined by

$$\mu_p\left(X_i^P\right)=f\left[D^w\left(X_i^P,C_p\right)\right]$$

  If $D^w\left(X_i^P,C_p\right)>\kappa$, then $\mu_p\left(X_i^P\right)=1$

  If $D^w\left(X_i^P,C_p\right)\le\kappa$, then $\mu_p\left(X_i^P\right)=1-\dfrac{D^w\left(X_i^P,C_p\right)}{\kappa}$,

  where κ is a predetermined overlapping threshold value.

Concluding Remarks

Fuzzy logic and neural networks were inspired by human computational powers. Both provide solutions that could be difficult to reach otherwise. However, the solutions reached may not be the optimum solutions to the problems. Combining the two techniques in a hybrid fuzzy and neural system enables enhanced performance. The hybrid system exploits the merits of each technique. Still, the solution reached may not be the optimum one. Genetic algorithms, GA (introduced in Appendix A), can be employed for optimization purposes in fuzzy logic systems, neural networks, and hybrid fuzzy-neural systems.

Bibliography

1. S. Abe and M.-S. Lan, "Efficient Methods for Fuzzy Rule Extraction from Numerical Data," in *Fuzzy Logic and Neural Network Handbook*, C. H. Chen Ed., McGraw Hill, New York, 1996, pp.7.1–7.33.

2. H. Adali and S.-L. Hung, *Machine Learning – Neural Networks, Genetic Algorithms and Fuzzy Systems*, John Wiley & Sons, New York, 1995.

3. C. von Altrock, *Fuzzy Logic and Neural Fuzzy Applications Explained*, Prentice hall PTR, Englewood Cliffs, NJ, 1995.

4. A. Blanco, M. Delgado and M.C. Pegalajar, "Identification of Fuzzy Dynamic Systems Using Max-Min Recurrent Neural Networks," *Fuzzy Sets and Systems*, 122, 3, 451– 467, 2001.

5. P. P. Bonissone, Y.-T. Chen, K. Goebel, and P. S. Khedkar, "Hybrid Soft Computing Systems: Industrial and Commercial Applications," *Proc. IEEE*, 87, 9, 1641–1667, 1999.

6. F. Bouslama and K. Ichikawa, "Application of Neural Networks to Fuzzy Control," *Neural Networks*, 6, 791–799, 1993.

7. J. J. Buckley and T. Feuring, *Fuzzy and Neural: Interactions and Applications*, Physica-Verlag, Heidelberg, 1999.

8. J. J. Buckley and Y. Hayashi, "Fuzzy Neural Networks: A Survey," *Fuzzy Sets and Systems*, 66, 1–13, 1994.

9. G. Carpenter and S. Grossberg, "Fuzzy ARTMAP: A Synthesis of Neural Networks and Fuzzy Logic for Supervised Categorization and Nonstationary," *Prediction, Fuzzy Sets, Neural Networks, and Soft Computing*, (R. R. Yager and L. A. Zadeh, eds.) Van Nostrand Reinhold, New York, 1994.

10. G. A. Carpenter, S. Grossberg, and D. B. Rosen, "Fuzzy ART: Fast Stable Learning and Categorization of Analog Patterns by Adaptive Resonance System," *Neural Networks*, 4, 759–771, 1991.

11. G. Castellano, A.M. Fanelli, and C. Mencar. "A Neuro-Fuzzy Network to Generate Human-understandable Knowledge from Data," *Cognitive Systems Research Journal*, 3, 2, 125–144, 2002.

12. G. Castellano, A.M. Fanelli and T. Roselli. "Fuzzy Rule Extraction by a KBN Approach," *First International ICSC Conference on Neuro-Fuzzy Technologies*, *NF 2002,* Havana, Cuba, January 16 – 19, 2002.

13. L. Caponetti, G. Castellano, and A.M. Fanelli, "A Neuro-Fuzzy System for Document Image Segmentation and Region Classification," *The 2nd IEEE International Workshop on Intelligent Signal Processing, WISP2001,* Budapest, Hungary, May 2001.

14. C. H. Chen, *Fuzzy Logic and Neural Network Handbook*, IEEE Press, New York, 1996.

15. J.-L. Chen and J.-Y. Chang, "Fuzzy Perceptron Neural Network for Classifiers with Numerical Data and Linguistic Rules as Inputs," *IEEE Trans. Fuzzy Systems*, 8, 6, 730–745, 2000.

16. C.-L. Chen and W.-C. Chen, "Fuzzy Controller Design by Using Neural Network Techniques," *IEEE Trans. Fuzzy Systems*, 2, 235–244, 1994.

17. M. Chiaberge and L. M. Reyneri, "Cintia: A Neuro-Fuzzy Real-Time Controller for Low-Power Embedded Systems," *IEEE Micro*, 6, 40–47, 1995.

18. F.-L. Chung and J.-C. Duan, "On Multistage Fuzzy Neural Network Modeling," *IEEE Trans. Fuzzy Systems*, 8, 2, 125–142, 2000.

19. F. L. Chung and T. Lee, "Fuzzy Competitive Learning," *Neural Networks*, 3, 539–551, 1994.

20. R. K. De, J. Basak, and S. K. Pal, "Unsupervised Feature Extraction Using Neuro-Fuzzy Approach," *Fuzzy Sets and Systems*, 126, 3, 277–291, 2002.

21. Y. Diano and K. M. Passino, "Adaptive Neural/Fuzzy Control for Interpolated Nonlinear Systems," *IEEE Trans. Fuzzy Systems*, 10, 5, 583–595, 2002.

22. J.-C. Duan and F.-L. Chung, "Cascaded Fuzzy Neural Network Model Based on Syllogistic Fuzzy Reasoning," *IEEE Trans. Fuzzy Systems*, 9, 2, 293–306, 2001.

23. M. J. Er and S. Wu, "A Fast Learning Algorithm for Parsimonious Fuzzy Neural Systems," *Fuzzy Sets and Systems*, 126, 3, 337–351, 2002.

24. M. French and E. Rogers, "Input/Output Stability Theory for Direct Neuro-Fuzzy Controllers," *IEEE Trans. Fuzzy Systems*, 6, 3, 331–345, 1998.

25. C. W. Frey and H.-B. Kuntze, "A Neuro-Fuzzy Supervisory Control System for Industrial Batch Processes," *IEEE Trans. Fuzzy Systems*, 9, 4, 570–577, 2001.

26. H. P. Graf and L. M. Reyneri, "The Expanding World of Neural and Fuzzy Systems," *IEEE Micro*, 6, 10–11, 1995.

27. M. M. Gupta and D. H. Rao, "On the Principles of Fuzzy Neural Networks," *Fuzzy Sets and Systems*, 61, 1–18, 1994.

28. K. Hirota and W. Pedrycz, "Neurocomputing with Fuzzy Flip-Flops," *Proceedings of the International Joint Conference on Neural Network*, 2, 1867–1870, 1993.

29. Q. Hu and D. B. Hertz, "Fuzzy Logic Controlled Neural Learning," *Information Sciences Applications*, 2, 1, 15–33, 1994.

30. H. Isshibuchi, R. Fujioka, and H. Tanaka, "Neural Networks that Learn from Fuzzy IF-Then Rules," *IEEE Transactions on Fuzzy Systems*, 1, 2, 85–97, 1993.

31. J.-S. R. Jang, C.-T. Sun, and E. Mizutani, *Neuro-Fuzzy and Soft Computing*, Prentice Hall, Uppwer Saddle River, NJ, 1997.

32. C.-F. Juang and C.-T. Lin, "An On-Line Self-Constructing Neural Fuzzy Inference Network and Its Applications," *IEEE Trans. Fuzzy Systems*, 6,1, 12–32, 1998.

33. S. V. Kartalopoulos, *Understanding Neural Networks and Fuzzy Logic*, IEEE Press, New York, 1996.

34. N. K. Kasabov, *Foundation of Neural Networks, Fuzzy Systems, and Knowledge Engineering*, The MIT Press, Cambridge, 1996.

35. T. Kasuba, "Simplified Fuzzy ARTMAP," *AI Expert*, 11, 18–55, 1993.

36. J. M. Keller and H. Tahani, "Implementation of Conjunctive and Disjunctive Fuzzy Logic Rules with Neural Networks," *International Journal of Approximate Reasoning*, 6, 221–240, 1992.

37. B. Kosko, *Neural Networks and Fuzzy Systems*, Prentice Hall, Englewood Cliffs, NJ, 1992.

38. S.-Y. Kung, J. Yaur, and S.-H. Lin, "Synergistic Modeling and Applications of Hierarchical Fuzzy Neural Networks," *Proc. IEEE*, 87, 9, 1550–1550–1574, 1999.

39. H. K. Kwan and Y. Cai, "Fuzzy Neural Network and Its Application to Pattern Recognition," *IEEE Trans. on Fuzzy Systems*, 2, 3, 185–193, 1994.

40. R. P. Landim, B. Rodrigues, S. R. Silva, and W. M. Matos, "A Neo-Fuzzy-Neuron with Real Time Training Applied to Flux Observer for an Induction Motor," *Proceedings of the 5th IEEE Brazilian Symposium on Neural Networks*, Belo Horizonte, MG, Brazil, December 9–11, 1998, pp.1–6.

41. R. P. Landim, B. Rodrigues, S. R. Silva, and W. M. Matos, "On-Line Neo-Fuzzy-Neuron State Observer," *Proceedings of the 6th IEEE Brazilian Symposium on Neural Networks*, Rio de Janeiro, Brazil, January 22–25, 2000, pp 196–201.

42. S. C. Lee and E. T. Lee, "Fuzzy Neural Networks," *Mathematical Biosciences*, 23, 151–177, 1975.

43. R.-P. Li, M. Mukaidono and I. B. Turksen, "A Fuzzy Neural Network for Pattern Classification and Feature Selection," *Fuzzy Sets and Systems*, 130, 1, 101–108, 2002.

44. C.-T. Lin, F.-B. Duh, and D.-J. Liu, "A Neural Fuzzy Network for Word Information Processing," *Fuzzy Sets and Systems*, 127, 1, 37–48, 2002.

45. C.-T. Lin and C. S. G. Lee, *Neural Fuzzy Systems*, Prentice Hall PTR, New York, 1996.

46. C.-J. Lin and C.-T. Lin, "An ART-Based Fuzzy Adaptive Learning Control Network," *IEEE Trans. Fuzzy Systems*, 5, 4, 477–496, 1997.

47. F.-J. Lin, C.-H. Lin, and P.-H. Shen, "Self-Constructing Fuzzy Neural Network Speed Controller for Permanent-Magnet Synchronous Motor Drive," *IEEE Trans. Fuzzy Systems,* 9, 5, 751–759, 2001.

48. Z.-Q. Liu and F. Yan, "Fuzzy Neural Network in Case-Based Diagnostic System," *IEEE Trans. Fuzzy Systems*, 5, 2, 209–222, 1997.

49. G.N. Marichal L. Acosta, L. Moreno, J.A. Méndez, J.J. Rodrigo, and M. Sigut, "Obstacle Avoidance for a Mobile Robot: A Neuro-Fuzzy Approach," *Fuzzy Sets and Systems*, 124, 2, 171–179, 2001.

50. J. Nie and D. Linkens, *Fuzzy-Neural Control: Principles, algorithms and applications*, Prentice Hall, New York, 1995.

51. S. K. Pal and S. Mitra, "Multilayer Perceptron, Fuzzy Sets, and Classifications," *IEEE Trans. on Neural Networks*, 3, 683–697, 1992.

52. W. Pedrycz, *Fuzzy Control and Fuzzy Systems*, John Wiley & Sons, New York, 1993.

53. V. B. Rao and H. V. Rao, *C++ Neural Networks & Fuzzy Logic*, MIS Press, New York, 1995.

54. P. K. Simpson, "Fuzzy Min-Max Neural Networks - Part 1: Classification," *IEEE Trans. on Neural Networks,* 3, 776–786, 1992.

55. P. K. Simpson, "Fuzzy Min-Max Neural Networks - Part 2: Clustering," *IEEE Trans. on Neural Networks*, 1, 32–45, 1993.

56. H. Takagi and I. Hayashi, "NN-Driven Fuzzy Reasoning," *International Journal of Approximate Reasoning*, 3,5, 191–212, 1991.

57. L. H. Tsoukalas and R. E. Uhrig, *Fuzzy and Neural Approaches in Engineering*, John Wiley & Sons, New York, 1997.

Web Resources

1. First IF/THEN for neural systems
   faculty.washington.edu/chudler/papy.html
   www.eoa.org.eg/edwintxt.htm#TOP

   These two sites provide a description of an ancient Egyptian papyrus that contains a surgery database in the form of an if/then rule. The papyrus has the first use of the neuro words recorded in history.

2. Neural Fuzzy Systems
   citeseer.nj.nec.com/64350.html

   A 1995, 250-page report by Robert Fullér on Neural Fuzzy Systems. The report consists of three chapters and an appendix. The first chapter discusses fuzzy systems, the second highlights basic concepts of neural networks, and the third introduces some aspects of fuzzy neural networks. The appendix provides a case study and exercise.

   Also,

   www.abo.fi/~rfuller/nfs.html

   Lecture notes on neural fuzzy systems in both Web and PDF formats by the same author.

3. Neural Networks and Fuzzy Systems
   http://nn.uidaho.edu/pap/2002\Mechatronics%20Handbook.pdf

   Chapter 32 authored by Bogdan Wilamowski in Mechatronics Handbook edited by Robert R. Bishop, CRC Press, pp. 33–1 to 32–26, 2002.

4. Evolving Fuzzy Neural Networks
   divcom.otago.ac.nz/infosci/kel/CBIIS/pubs/pdf/ajiips-si98.pdf

   A paper that appeared in the Australian Journal of Intelligent Information Processing Systems, 5,154-160,1998 by Nikola Kasabov, University of Otage, New Zealand. It is a 6-page paper with 29 references available in PDF format. The complete title is Evolving Fuzzy Neural Networks: Theory and Applications for On-line Adaptive Prediction, Decision Making in Control.

   citeseer.nj.nec.com/kasabov01evolving.html

   A 67-page paper with 84 references by the same author. The full title is Evolving Fuzzy Neural Networks for Supervised/unsupervised On-line Knowledge-Based Learning. It appeared in IEEE Transactions of Systems Man and Cybertics, 131, 6, 2001.

5. Adaptive Neuro-Fuzzy Inference System, ANFIS
    www.control.hut.fi/Kurssit/AS-74.115/Material/FVAnfis2.pdf

    A 25-page tutorial by Heikki Koivo on the Adaptive Neuro-Fuzzy Inference System. It has MATLAB examples, copyrighted 2000. Also, lecture notes on Neuro Fuzzy Computing in Automation by Prof. Koivo, 2001, available at www.control.hut.fi/Kurssit/AS-74.115/Material/

6. Neural Fuzzy Motion Estimation and Compensation
    sipi.usc.edu/~kosko/MotionEstimation.pdf

    A paper by Hyum Kim and Bart Kosko, IEEE Transactions on Signal Processing, 45, 10, 2515-2532, 1997. Provided in PDF format. The paper discusses the use of neural-fuzzy systems to improve motion estimation and compensation for video compression, and it has 29 references.

7. Soft Negotiation
    www.iis.ee.ic.ac.uk/~frank/surp98/report/mgm1/

    A report in Web format by Myles MacRae and Marcus Pickering, the Department of Computing, Imperial College of Science, Technology and Medicine, University of London. The report covers several topics including a simple, short account of soft computing. The account addresses neural networks, fuzzy logic, genetic algorithms and their combinations.

8. Neural Fuzzy Techniques
    scholar.lib.vt.edu/theses/available/etd-5733142539751141/unrestricted/
    ETD.PDF

    A 187-page Ph.D. dissertation by Somkiat Sampan, Virginia Polytechnical Institute and State University, 1997. It is provided in PDF format. The dissertation discusses neural fuzzy techniques in vehicle acoustic signal classification.

9. Neural Fuzzy Inference Network
    www.di.uniba.it/~castella/papers/IJCNN2000.pdf

    A paper by Castellano and Fanelli, Università degli Studi di Bari, Italy. It presents a self-organizing neural fuzzy inference network. It is 6 pages long, with 13 references. It is available in PDF format.

10. Fuzzy ARTMAP vs. MLP
    www.cairo.utm.my/publications/yhtay_airtc97.pdf

    A paper available in PDF format: Tay and Kalid, Comparison of fuzzy ARTMAP and MLP Neural Networks for Handwritten Character Recognition, IFAC Symposium on AI in Real-Time Control, Kuala Lumpur, Malaysia, 1997.

The paper presents results of comparisons between Fuzzy ARTMAP and backpropagation based multilayer perceptrons, MLP. Fuzzy ARTMAP out performed MLP in both learning convergence and recognition accuracy.

11. Fuzzy ARTMAP Applications
    www.cairo.utm.my/publications/ksyap_wec99.pdf

    Slah et al., Vehicle Licence Plate Recognition by Fuzzy ARTMAP Neural Network, World Engineering Congress, WEC'99, Universiti Putra Malaysia, 1999. It is a six page report with seven references available in PDF format. It reports on the prototype of a system being developed at the Center for Artifical Intelligence and Robotics in Cairo and the Universiti Teknologi Malaysia.

12. Simplified Fuzzy ARTMAP Applications medusa.sdsu.edu/Robotics/
    Neuromuscular%20Control/Fuzzy_ARTMAP.pdf

    A paper by Sajasekarn and Pai, PSG College of Technology titled: Image Recognition using Simplified Fuzzy ARTMAP, 15 pages, 12 references.

13. Combining Neural Networks and Fuzzy Controllers
    citeseer.nj.nec.com/nauck93combining.html

    A paper by Nauck, Klawonn, and Kruse that appeared in Fuzzy Logic in Artificial Intelligence, FLAI93; it has 13 pages and 21 references, and is available in PDF format.

14. Modified Fuzzy Neural Network Classifier
    www.ee.iitb.ac.in/uma/~ncc2002/proc/NCC-2002/pdf/n010.pdf

    This paper discusses neural networks based on modifying the Kwan-Cai Networks.

    Authored by Kularni and Sontakke, SGGS College of Engineering and Technology, India.

15. Comparative Study of NN Structures
    mecha.ee.boun.edu.tr/~efe/PDF/JMechatronics.pdf

    Efe and Kayank of Bogazci Univesity, Turkey present a 16-page report that presents the results of investigating the identification of nonlinear systems by neural networks. Feedforward Neural Networks, Radial Basis Function Neural Networks, Rung-Kutta Neural Networks, and Adaptive Neuro Fuzzy Inference Systems are evaluated and their performance compared.

16. Soft Computing Resources
    www.cs.nthu.edu.tw/~jang/nfsc.htm

    A web site maintained by Prof. Jang, Computer Science Department, Tsing Hua University, Taiwan. It provides Web links to numerous resources including FTP sites, newsgroups, neurofuzzy research sites, journals, technical reports, and more.

17. Computational Intelligence Links
    www.emsl.pnl.gov:2080/proj/neuron/ci/what.html

    This Web sites provides a large number of Web links to mainly introductory information on Computational Intelligence (Neural, Fuzzy, and Genetic). It is maintained by Pacific Northwest National Laboratory, operated by Battelle, an industry located in Ohio, USA.

18. Neuro Fuzzy Resources
    elve.le.ttu.ee/mesel_www_home/R&D/NEUROFUZ/Ressourc.htm

    A large collection of Web links to resources including tutorials, courses, journals, commercial companies, and much more. It is maintained by Martin Brown, Department of Electronics, Tallinn Technical University, Estonia.

19. Neuro Fuzzy Systems: State-of-the-art Modeling Techniques
    citeseer.nj.nec.com/abraham01neuro.html

    A document from Lecture Notes in Computer Science by Ajith Abraham of Monash University, Australia. It is an eight page paper with eleven references that suggest a modeling approach for neuro-fuzzy systems.

CHAPTER **8**

Hardware Implementations

8.1 Introduction

Embedded systems and their realizations were discussed in Chapter 1. The general features, design metrics, processor technologies, and the IC technologies which were discussed in general terms for embedded systems are applicable to embedded fuzzy and neural systems. Specifically, the approaches that have been used to implement embedded fuzzy logic systems and neural networks include:

- Using general-purpose processors and depending fully on software in the realization of the system. As discussed earlier, such an approach is relatively easy and low-cost, but also leads to relatively slow operation.

- Adapting a general-purpose processor to perform dedicated fuzzy instructions. The approach is a trade-off between speed and generality. Using a coprocessor or exclusive hardware to perform the fuzzy operations is a closely related approach, and the speed is limited by the interfacing between the processor and the co-processor.

- Using dedicated fuzzy circuits or Application Specific Integrated Circuits, (ASICs). The approach leads to relatively high-speed operation, but is more costly.

In Chapter 9, we will discuss software tools, including those for the implementation of fuzzy and neural systems on general-purpose processors. Also discussed are software tools that enable the conversion of a given intuitively designed system to code in a selected language such as C, C++, Java, etc., or to an assembly language of a particular general-purpose, or dedicated, processor. The following sections discuss the hardware aspect of the embedded-system implementation. The techniques used for hardware implementation include:

- Digital techniques
- Continuous-time analog techniques including:
  - Current mode circuits
  - Voltage mode circuits
  - Mixed mode (current and voltage) circuits

■ Discrete-time analog techniques

■ Mixed signal (analog and digital) techniques

The discussion is meant to give the reader an overview of the basic ideas of hardware implementation using digital and analog techniques. Since new ICs are proposed, designed, and produced at a rapid pace, the reader is advised to refer to the relevant parts of the Web resources section if specific up-to-date information is needed.

8.2 Digital Techniques

Although fuzzy chips may have limited input/output capabilities, they have particularly useful applications in real-time control systems. Analog fuzzy values must be converted to binary digital signals. Analog-to-digital conversion can lead to quantization errors in both input signals and membership values. Thus, a deterioration in the fuzzy processing may occur if an insufficient number of bits is used to represent the analog signals. On the other hand, using a large number of bits can slow down the process. This is the trade-off between precision and speed.

Fuzzy dedicated circuits are characterized by:

■ The number of inputs and outputs.

■ The number and shapes of membership functions.

■ Inference techniques including operators, consequences, and size of the premises.

■ Defuzzification method.

■ The number of fuzzy logic inferences per second, FLIPS.

■ Physical size.

■ Power consumption.

■ Software available to support the design.

Numerous approaches to digital fuzzy ICs have been reported.

Example

First Digital Fuzzy IC

Togai et al. proposed the first fuzzy logic digital integrated circuit implementation in the mid-1980s. In the first prototype, they emphasized simplicity, making it particularly useful as a starting point in understanding custom digital IC implementation of fuzzy logic inference. The implementation is based on the analysis summarized here.

A fuzzy rule is defined by a relation between observation (or antecedent) and action (or conclusion). For a given set of fuzzy rules, the action is inferred from observation and a fuzzy relation. Now, suppose $A_1, A_2, \ldots, A_N$, are fuzzy subsets of a universe of discourse U, and $B_1, B_2, \ldots, B_N$ are fuzzy subsets of the universe of discourse V. A fuzzy relation is defined by rules such as

IF A_1 THEN B_1,

IF A_2 THEN B_2,

.

.

.

IF A_N THEN B_N.

All the rules are combined to give an overall fuzzy relation R. The fuzzy relation R_i is constructed from Rule i and linguistic values A_i and B_i.

Consider the ith rule of a set of N rules. Given an observation A' and a rule R_i, the action B_i' is inferred by:

$$B_i' = A' \circ R_i ; \qquad A' \in U, \; B_i' \in V, \text{ and } R_i \subset U \times V$$

$$\mu_{B_i'}(v) = \max \min \left(\mu_{R_i}(u), \mu_{A_i}(u,v) \right)$$

$$= \min \max \left[\min \left(\mu_{A'}(u), \mu_{A_i}(u) \right), \mu_{B_i}(v) \right]$$

$$= \min \left(\alpha_i, \mu_{B_i}(v) \right)$$

where $\alpha_i = \max \min \left(\mu_{A'}(u), \mu_{A_i}(u) \right)$

Then, the maximum of $B_1, B_2, \ldots, B_N$ determines the overall resulting decision (action) B', i.e. $B' = \bigcup_i B_i'$.

The inference mechanism analyzed lends itself to VLSI realization by a logical architecture with two-level hierarchy using min and max operations. Ordinary OR and AND gates were used and a custom CMOS technology was employed. The prototype implemented 16 rules with one antecedent and one consequent. It represented each fuzzy label by 31 elements with 16 possible values of membership. The maximum inference speed was 80,000 FLIPS.

8.3 Analog Techniques

The variables in fuzzy and neural systems are analog by nature. Thus, analog implementation eliminates the need for analog-to-digital and digital-to-analog conversions. The systems also require massive parallelism, making analog circuits particularly suited for their implementation. Further, the physical characteristics of transistors can be utilized in realizing the nonlinear functions required, whether it is a fuzzy operation, a membership function, or a neuron activation function. Analog implementations, however, have typically very restricted possibilities for programmability. Analog implementation techniques include voltage mode and current mode realizations, in addition to mixed mode (current and voltage) realizations.

8.3.1 Voltage mode

VLSI voltage-mode implementation has the merit of the simplicity of signal voltage distribution in various parts of the circuit. However, the signals in the circuit are sensitive to changes in supply voltages. Further, charge delays of various parasitic capacitances limit the speed. Numerous circuits and architectures have been suggested and implemented; some are listed in the bibliography and the Web resources section. The following example illustrates the concept of repressing fuzzy operations using hardware by giving simple circuit examples of fuzzy INVERTER, AND, and OR operations.

Example

Fuzzy INVERTER

The standard fuzzy complement was defined in Chapter 2 by:

$$\mu_{\bar{A}}(x) = 1 - \mu_A(x)$$

This equation can be translated into a circuit with an input-output reaction given by:

$$V_o = V_{ref} - V_i$$

where V_{ref} and V_i are a reference and input voltage, respectively.

The input-output relation can be expressed normalized, dividing by V_{ref}, as:

$$V_o = 1 - V_i$$

Thus, $\mu_{\bar{A}}(x)$ is represented by V_o and $\mu_A(x)$ is represented by V_i. The circuit shown in Figure 8.1 can achieve the required input-output relation.

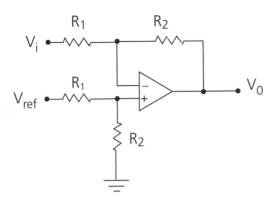

Figure 8.1: A circuit to illustrate fuzzy INVERTER hardware implementation.

In the circuit of Figure 8.1, we can write:

$$V_o = \left(\frac{R_2}{R_1 + R_2}\right)\left(\frac{R_1 + R_2}{R_1}\right)V_{ref} - \left(\frac{R_2}{R_1}\right)V_i$$

This leads to:

$$V_o = \left(\frac{R_2}{R_1}\right)\left(V_{ref} - V_i\right)$$

Then, with $V_{ref} = 1$, we obtain:

$$V_o = K\left(1 - V_i\right)$$

where K is a constant.

Fuzzy OR

The standard fuzzy OR operation was also defined in Chapter 1 by:

$$\mu_{A \cup B}(x) = \max\left[\mu_A(x), \mu_B(x)\right]$$

This function can be realized by a circuit that has an output voltage, V_o, equal to the value of the largest of the two input voltages. The circuit shown in Figure 8.2 is a generalization from a simple realization of a Boolean OR.

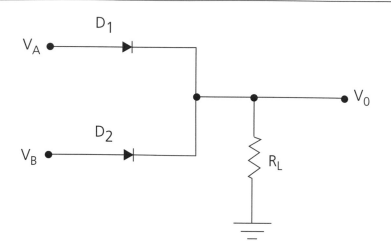

Figure 8.2: A circuit to illustrate fuzzy OR hardware implementation, where D$_1$ and D$_2$ represent precision rectifier circuits.

In the circuit of Figure 8.2, the output voltage, V_o, is given by:

$$V_o = V_A \text{ , if } V_A > V_B$$

$$V_o = V_A, \text{ if } V_B > V_A$$

In Boolean logic realization, the inputs and output may assume one of two possible values—e.g., 0V and 5V. Diodes can be used without considering the voltage needed to turn on the diode, V_D, since it is about 0.7V. In fuzzy logic, the inputs and outputs may assume any value in the range 0V to 5V, or normalized range 0 to 1. A diode with $V_D = 0$ (or close to zero) would be needed. This is the reason for designating D_1 and D_2 as precision rectifiers in Figure 8.2.

Fuzzy AND

An expression for a fuzzy AND operation can be expressed using OR and INVERTER operations through the applications of DeMorgan's theorem as follows:

$$A \cdot B = \left(A' + B'\right)'$$

Thus, a fuzzy AND can be implemented using inverting precision rectifier and INVERTER circuits.

Pérez and Bañuelos presented the details and experimental results of implementing basic fuzzy logic gates along similar lines to the above discussion (see *Electronic Models of Fuzzy Gates* in the Web Resources section at the end of this chapter).

8.3.2 Current mode

This technique is becoming more and more popular in VLSI implementations because of its numerous advantages. Current mode circuits do not require resistors in their implementations. This is advantageous because integrated resistors are typically inaccurate and involve a significant amount of parasitic capacitance. Furthermore, numerous operations, including summation and subtraction, can be achieved easily using current mode circuits. In addition, these circuits exhibit low sensitivity to supply voltage variations and have lower power consumption compared to the voltage mode circuits.

In the following discussion, some basic concepts and building blocks are presented. These should help visuale how fuzzy circuits can be implemented in current mode. They should also help the reader to follow and appreciate the numerous implementations cited at the end of the chapter.

8.3.2.1 Current mirrors

Current mirror circuits are used to distribute the signal current to various parts of the circuit. They produce a copy, or copies, of a given current signal. They are used in the design of current sources for biasing as well as operating as active loads. Numerous current mirror circuit configurations that been suggested and implemented in both MOS and bipolar technologies.

Example

Bipolar Junction Transistor, BJT, Current Mirror Circuit

The circuit shown in Figure 8.3 acts as a current mirror. It copies an input, or reference current, I_{ref}, to an output current, I_o. It uses two matching transistors Q_1 and Q_2 which implies that they have the same v_{BE}. Transistor Q_1 is connected as a diode by short-circuiting its collector to its base. In the circuit shown:

$$I_o = \frac{\beta}{\beta+1} I_E \text{, and}$$

$$I_{ref} = \frac{\beta+2}{\beta+1} I_E$$

leading to:

$$I_o = \frac{\beta}{\beta+2} I_{ref} = \frac{1}{1+\frac{2}{\beta}} I_{ref}$$

Then, for $\beta \gg 1$, $I_o = I_{ref}$

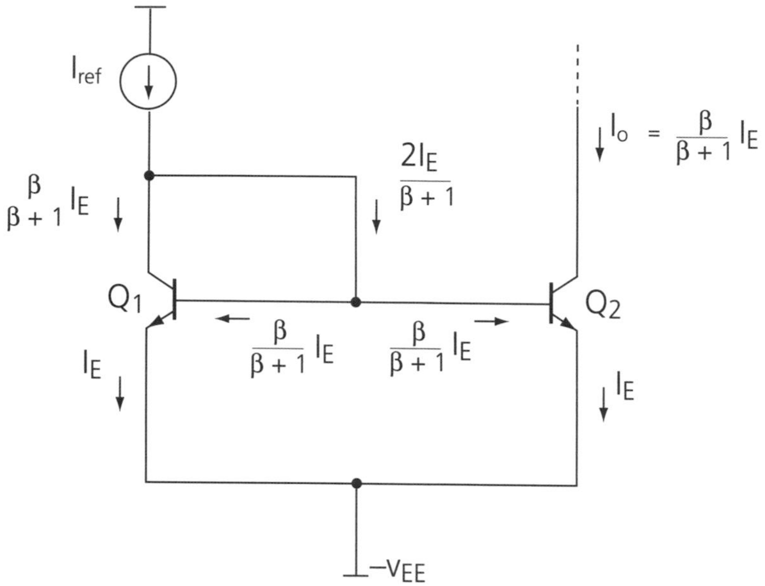

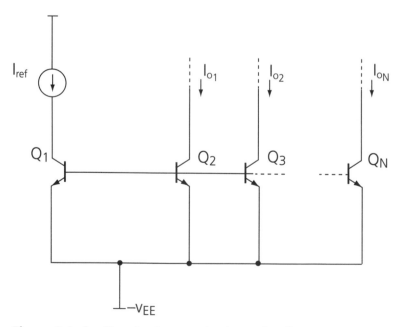

Figure 8.3: A basic BJT current mirror circuit.

The circuit discussed in this example can be generalized to become an *N*-output current mirror as shown in Figure 8.4.

Figure 8.4: An *N*-output current mirror circuit.

Example

A Basic MOSFET Current Mirror Circuit

The basic MOSFET current mirror consists of two enhancement MOSFETs as shown in Figure 8.5. For Q_1, the input (reference) current is given by:

$$I_{ref} = K_1 (V_{GS} - V_{th})^2$$

The output current is given by:

$$I_o = K_2 (V_{GS} - V_{th})^2$$

where K_1 and K_2 are constants that depend on the geometry of the devices (the ratio of the width of the channel to its length), and V_{th} is the threshold voltage.

The above equations lead to:

$$I_o = KI_{ref} \text{ , where } K = \frac{K_1}{K_2}$$

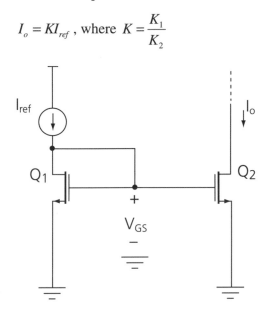

Figure 8.5: A basic MOSFET current mirror circuit.

Example

Improved MOSFET Current Mirror Circuit

An ideal current source has an infinite output resistance and the output current is independent of the output voltage. Thus, increasing the output resistance of a current mirror improves its performance. This could be achieved using the so-called cascode mirror (stacked mirror) shown in Figure 8.6.

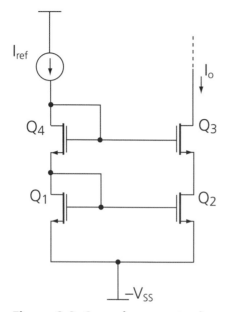

Figure 8.6: Cascode current mirror.

The output resistance of the basic current mirror is r_{o2} which is the output resistance of transistor Q_2. The output resistance of the cascode mirror can be calculated using the equivalent circuit shown in Figure 8.7.

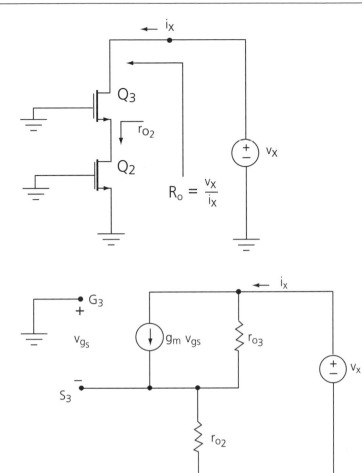

Figure 8.7: Calculating the output resistance of the cascode current mirror.

The output resistance is given by

$$\frac{v_x}{i_x} = r_{o3} + r_{o2} + g_m r_{o3} r_{o2}$$

$$\approx (g_{m3} r_{o3}) r_{o2}$$

Thus, the output resistance in the cascode configuration is higher by a factor of $(g_m r_{o3})$ over that of the basic current mirror.

Example

The Wilson Current Mirror Circuit

Another current mirror circuit configuration, referred to as the Wilson current mirror, is shown in Figure 8.8. This configuration leads to increased output resistance as in the cascode configuration. However, the drain voltages of the two transistors are not equal, and hence their currents are not equal either. The problem is overcome by the so-called Modified-Wilson circuit, where a diode-connected transistor Q_4 as shown in Figure 8.9.

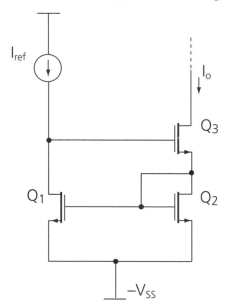

Figure 8.8: Wilson current mirror.

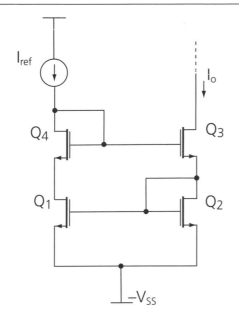

Figure 8.9: Modified-Wilson current mirror.

Example

BiCMOS Double Cascode Current Mirror

The current mirror circuit shown in Figure 8.10 is referred to as a double cascode BiCMOS. It uses both bipolar and MOSFET transistors. The output resistance of the circuit is given by $R_o = (g_{m3}r_{o3})(\beta_2 r_{o2})$, indicating that this current mirror has extremely high output resistance.

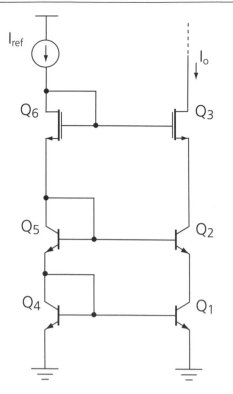

Figure 8.10: The circuit of a BiCMOS double cascode current mirror.

8.3.2.2 Fuzzy Operators

$$I_Z = \begin{cases} I_A - I_B, I_A \geq I_B \\ 0, I_A < I_B \end{cases}$$

Numerous bipolar and MOSFET-based circuits have been devised to implement various fuzzy operations. The following are examples of circuit realizations of some basic fuzzy logic functions.

Bounded Difference

The bounded difference operation may be defined by:

$$\mu_Z = \mu_A \ominus \mu_B = \begin{cases} \mu_A - \mu_B, \mu_A \geq \mu_B \\ 0, \mu_A < \mu_B \end{cases}$$

A bounded difference circuit suggested by Yamakawa is shown in Figure 8.11. The circuit consists of a current mirror and a diode. The currents I_A, I_B, and I_Z represent the membership values μ_A, μ_B and μ_Z, respectively. The current I_B is applied to the current mirror, and the current I_B is wire-subtracted. Thus, the current at the anode of the anode of the diode is $(I_A + I_B)$. The output current , I_Z, will be:

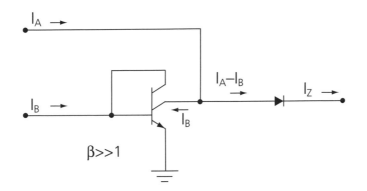
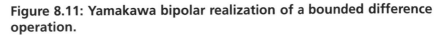

Figure 8.11: Yamakawa bipolar realization of a bounded difference operation.

The bounded difference operation can also be realized using MOSFETs as shown in Figure 8.12.

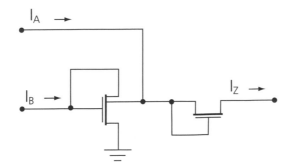
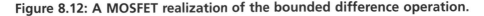

Figure 8.12: A MOSFET realization of the bounded difference operation.

If the circuit is cascaded by a current mirror, the diode-connected transistor can be omitted. The input of the current mirror of the next stage achieve the task of the diode.

Fuzzy Inverter Circuit

A basic fuzzy complement operation is defined by

$$\mu_Z = \mu_{\bar{A}} = 1 - \mu_A$$

The circuit shown in Figure 8.13 shows a possible realization of that operation.

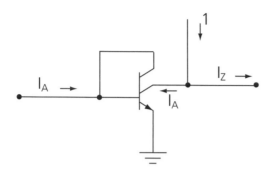

Figure 8.13: A possible bipolar realization of a fuzzy inverter; the currents represent the membership values.

Fuzzy AND (MIN circuit)

The fuzzy AND operation can be defined as

$$\mu_Z = \mu_{A \cap B} = \begin{cases} \mu_B, & \mu_A \geq \mu_B \\ \mu_A, & \mu_A < \mu_B \end{cases}$$

$$= \mu_B \ominus (\mu_B \ominus \mu_A)$$

$$= \mu_A \ominus (\mu_A \ominus \mu_B)$$

As before, the membership values are represented by current values and hence the above description can be realized by cascading two bounded-difference circuits as shown in Figure 8.14. The diode of the first stage is omitted as mentioned earlier, since the input characteristics of the following stage has rectification characteristics.

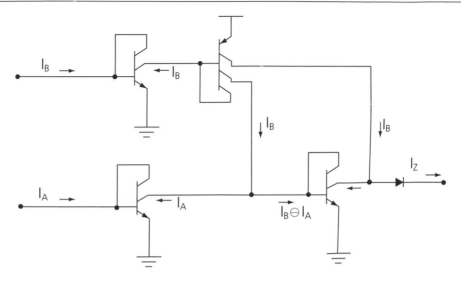

Figure 8.14: A bipolar realization of a fuzzy AND.

Fuzzy Logic OR (MAX circuit)

The fuzzy logic OR may be expressed as:

$$\mu_Z = \mu_{A \cup B} = \begin{cases} \mu_A, \mu_A \geq \mu_B \\ \mu_B, \mu_A < \mu_B \end{cases}$$

$$= (\mu_A \oplus \mu_B) + \mu_B$$

$$= (\mu_B \oplus \mu_A) + \mu_A$$

This description leads to the circuit realization shown in Figure 8.15.

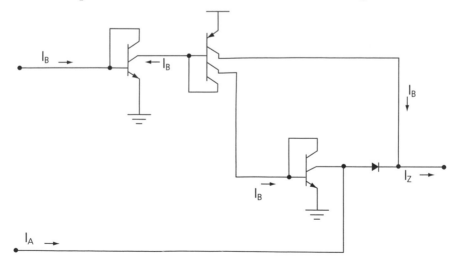

Figure 8.15: A bipolar realization of a fuzzy OR.

8.3.3 Mixed Mode

Both current-mode and voltage-mode circuits can be implemented on the same chip in a hybrid configuration. Such a configuration takes advantage of the merits of the two circuit types. A linear and accurate conversion between the two modes is essential in such implementations. One possible approach is to use Operational Transconductance Amplifiers, OTAs. A differential input OTA is shown in Figure 8.16 along with its circuit implementation. The output current, I_o, is related to the input voltages by:

$$I_o = g_m \left(V_1 - V_2 \right)$$

where g_m is a gain parameter that represents the transfer conductance of the amplifier (referred to as transconductance).

An ideal transconductance amplifier has an infinite input resistance and an infinite output resistance.

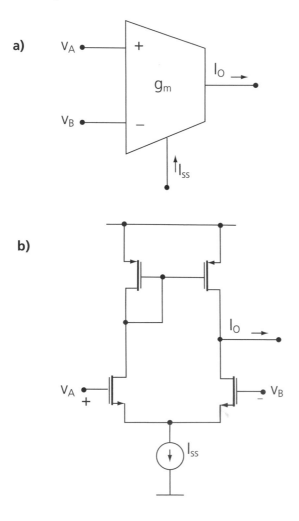

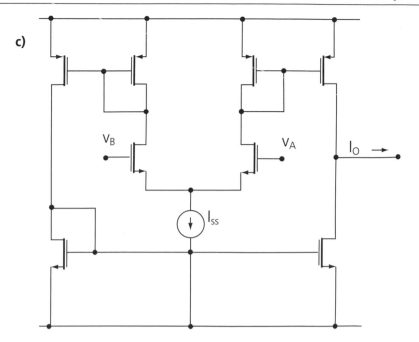

Figure 8.16: An OTA and some possible implementations.
   a) Circuit Symbol.
   b) Circuit of a simple differential input OTA.
   c) Circuit of a balanced OTA.

An OTA-based bounded difference circuit is shown in Figure 8.17. The first OTA has an output current of $g_m(V_A - V_B)$ which realizes the bounded difference for $\mu_A \geq \mu_B$, with the voltages V_A and V_B representing the membership functions μ_A and μ_B. The unidirectional characteristic of the diode enforces the condition $\mu_A < \mu_B$ as well. The second OTA converts the current at its input into a proportional output voltage, leading to a voltage-mode bounded difference.

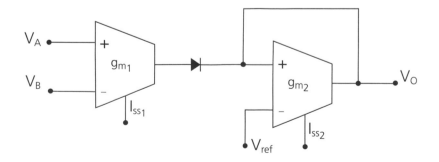

Figure 8.17: OTA-based bounded difference.

Once a bounded difference operation is realized, all fuzzy logic functions can be implemented using the bounded-difference circuit as a building block. Examples of fuzzy AND, fuzzy OR, and fuzzy NOT circuit realizations are shown in Figures 8.18, 8.19, and 8.20.

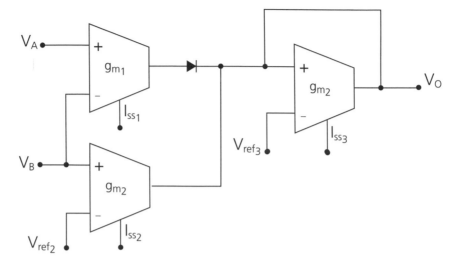

Figure 8.18: An OTA-based fuzzy AND circuit.

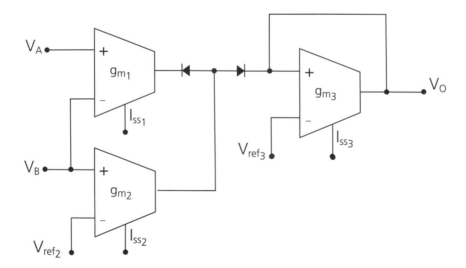

Figure 8.19: An OTA-based fuzzy OR circuit.

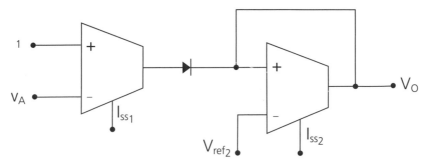

Figure 8.20: An OTA-based fuzzy INVERTER circuit.

8.4 Fuzzy Analog Memory

One difficult aspect of analog circuit implementations is devising a reliable analog memory module. There is no long-term accurate way to store analog fuzzy values. Programmability and multistage sequential processing are thus relatively difficult tasks in analog fuzzy processors. Hirota et al. have carried out extensive studies presenting the concept of fuzzy logic flip-flops, then implementing them along with fuzzy registers.

Example

Hirota et al. extended the binary J-K flip-flop to a fuzzy flip-flop, the operation of which is described by:

$$Q(t+1) = \{J \vee (1-K)\} \wedge (J \vee Q) \wedge \{(1-K) \vee (1-K)\}$$

Based on the above description, a fuzzy flip-flop was implemented using fuzzy gates (min, max, and inverter circuits).

Example

Hirota et al. explained that by using a fuzzy flip-flop circuit, one element of fuzzy information can be stored. Thus, a membership function can be stored using N fuzzy flip-flops and N fuzzy AND gates as illustrated in Figure 8.21. In that circuit, the membership value $\mu_A(x)$ is defined for $x_1, x_2, \ldots, x_N$. The operations of the control inputs are summarized as follows:

	State		
Control Input	Reset	Write	Shift
Reset	1	0	irrelevant
Shift	0	0	1

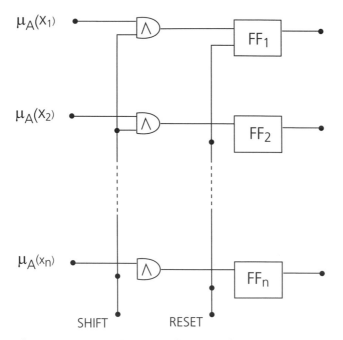

Figure 8.21: Storing membership functions using fuzzy flip-flops.

8.5 Neurons

The discussion in the previous sections focused on the implementation of fuzzy logic functions. Neural networks follow a similar path in their hardware implementations. Digital and analog implementations including voltage and current mode are employed. A hardware implementation of a neural network would be ideal if it works for any training algorithm, provides global interconnection with a large number of neurons, and has programmable weights, in addition to modular structure, low power consumption, and low price.

In this section only the concept of implementation of the neuron model using operational amplifier based circuits is presented as an example of analog implementation.

Figure 8.22 shows a zero-bias configuration of the basic neuron model. The output, Z, is given by:

$$Z = \sum_{i=1}^{n} x_i \left(-R_f G_i \right)$$

where $G_i = \dfrac{1}{R_i}$,

$x_1, x_2, \ldots, x_n$ are inputs, with the associated weights w_i given by $w_i = -R_f G_i$.

A bias can be introduced by connecting the non-inverting input terminal of the op-amp to a bias voltage, or fixing one of the input voltages to a constant value. Positive and negative weights can be implemented as shown in the circuit of Figure 8.23.

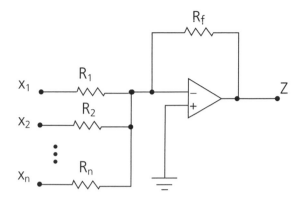

Figure 8.22: An op-amp based implementation of the basic neuron model.

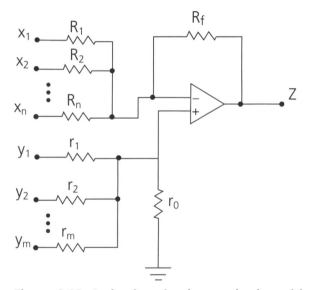

Figure 8.23: A circuit to implement both positive and negative weights.

Although the circuits of Figures 8.22 and 8.23 illustrate the basic concept of neuron implementation, they are of limited practical significance. Such neurons cannot learn, because the weights are fixed. Further, the modification possibilities of the transfer function characteristics are limited. Also, the circuit size and its power consumption are relatively large.

MOSFETs can be used as voltage-controlled resistors to achieve weight controllability as illustrated in Figure 8.24. The resistance of the channel, R_{ds}, between the drain and the source of each transistor is controlled by the voltage between the gate and the source, V_{gs}.

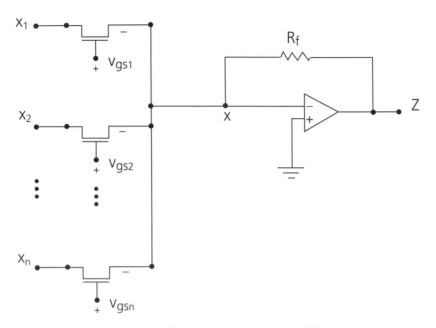

Figure 8.24: Neuron model implementation with voltage-controlled weights.

Weights can be stored using a capacitive circuit element. The elementary circuit shown in Figure 8.25 illustrates the concept. A weight could be stored as a voltage across the gate capacitance, since no current is needed to drive the gate. A capacitance could be added to the circuit to enhance the storage capability of the intrinsic gate capacitance. The weight control voltage, v_c, is sampled through transistor $Q_{S/H}$ which provides the action of an on/off switch. The resistance of transistor Q_1 is controlled by the voltage V_{gs1}, as indicated earlier. The weights could, of course, be stored digitally, in which case an analog-to-digital conversion would be required. More elaborate implementations have been put forward by numerous researchers; for example see the extensive work reported by Wilamowski et al. included in the Web resources.

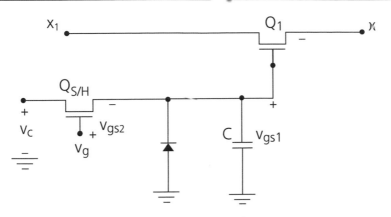

Figure 8.25: Elementary analog weight storage.

Concluding Remarks

The increasing interest in embedded systems applications of fuzzy logic has led to a growing interest in VLSI implementations of fuzzy logic functions, controllers, and systems in general. Both analog and digital implementations have received attention; each has its merits and demerits. Hybrid (analog/digital) systems have proven to be useful as well. Researchers have put forward numerous techniques for analog implementations along with their hybridizations. The discussion in this chapter was meant to introduce the reader to the fundamentals of a wide spectrum of hardware implementation techniques. This information provides the basic building blocks that appear, as is or modified, in various VLSI realizations of fuzzy systems. The reader may consult the available literature for implementation details of a specific system. Fundamental circuits for VLSI implementation of neural models were presented as well, the objective being the illustration of the way simple neuron models can be implemented and to facilitate further exploration in that area.

Bibliography

General

1. I. Batroune, A. Barriaga, S. Sánchez-Solano, G. J. Jiménez-Fernández, and D. R. López, *Microelectronic Design of Fuzzy Logic-Based Systems*, CRC Proess, New York, 2000.

2. P R. Gray and R. G. Mayer, *Analysis and Design of Analog Integrated Circuits*, John Wiley & Sons, Toronto, 1984.

3. M. M. Gupta and T. Yamakawa, Editors, *Fuzzy Computing: Theory, Hardware, and Applications,* North-Holand, 1988.

4. A. M. Ibrahim, *Introduction to Applied Fuzzy Electronics*, Prentice Hall, Upper Saddle River, NJ, 1997.

5. M. Jamshidi, N. Vadiee, T. Rose, *Fuzzy Logic and Control: Software and Hardware Applications*, Prentice Hall PTR/Sun Microsystems Press,1998.

6. A. Kandel and G. Langholz, Editors, *Fuzzy Hardware: Architectures and Applications*, Kluwer Academic Publishers, Boston, 1998.

7. C. Mead, *Analog VLSI and Neural Systems*, Addison-Wesley, Don Mills, Ontario, 1989.

8. M. J. Patyra and D. M. Mlynek, Editors, *Fuzzy Logic: Implementation and Applications,* Wiley/Teubner, New York, 1996.

9. W. Pedrycz, *Fuzzy Sets Engineering*, CRC Press, Boca Raton, Florida, USA.

10. A. S. Sidra and K. C. Smith, *Microelectronic Circuits*, Saunders College Publishing, Toronto, 1989.

11. P. K. Simpson, Editor, *Neural Networks Theory, Technology, and Applications*, IEEE Press, New York, 1996.

Digital Techniques

1. M. Chiaberge, E. Miranda, and L. M. Reyneri, "An HW/SW Co-Design Approach for Neuro-Fuzzy Hardware Design," *Proceedings of the 7ᵗʰ International Conference on Microelectronics for Neural, Fuzzy, and Bio-Inspired Systems*, Granada, Spain, April 7–9, 1999, pp. 332–337.

2. A. Costa, A. De Gloria, F. Giudici, and M. Olivieri, "Fuzzy Logic Microcontroller," *IEEE Micro*, 17, 1, 66–74, 1997.

3. H. Eichfeld, T. Kunemund, and M. Menke, "A 12 Bit General Purpose Fuzzy Logic Controller Chip," *IEEE Transactions on Fuzzy Systems*, 4, 4, 460–475, 1996.

4. R. Lacoste, "A Low-cost Neural Network Processor," *Circuit Cellar*, 114, 26–35, 2000.

5. M J. Patyra, J. L. Granter, and K. Koster, "Digital Fuzzy Logic Controller: Design and Implementation," *IEEE transactions on Fuzzy Systems*, 4, 4, 439–459, 1996.

6. R. Rovatli, "High Speed Implementation of Piecewise-quadratic Togai-Sugeno Systems with Memory Savings," *Proceedings of the World Multi-Conference on Circuits, Systems, Communications and Computers*, Athens, Greece, 1999, pp. 6451–6454.

7. R. Rovatti and M. Vitluari, "Linear and Fuzzy Piecewise-linear Signal Processing with Extended DSP Architecture," *Proceedings of the IEEE International Conference on Fuzzy Systems,* Anchorage, Alaska, USA, 1998, pp. 1082–1087.

8. V. Salapura, "A Fuzzy RISC Processor," *IEEE Transactions on Fuzzy Systems*, 8, 6, 781–790, 2000.

9. M. Sasaki, F. Ueno, and T. Inoue, "7.5 MFLIPS Fuzzy Processor Using SIMD and Logic-in-Memory Structure, *Proceedings of the IEEE International Conference on Fuzzy Systems*, San Francisco, California, USA, 1993, 527–534.

10. M. Togai and H. Watanabe, "A VLSI Implementation of Fuzzy-Inference Engine: Toward an Expert System on a Chip," *Information Sciences*, 38, 147–163, 1986.

11. H. Watanabe and D. Chen, "Evaluation of Fuzzy Instructions in a RISC Processor," *Proceedings of the IEEE International Conference on Fuzzy Systems*, San Francisco, California, USA, 1993, pp. 521–526.

Analog Techniques

1. F. Balteanu, I. Opris, and G. Kovacs, "Current-Mode Fuzzy Memory Element," *Electronics Letters*, 29, 2, 236–237, 1993.

2. C. Giustolisi, G. Palumbo, and S. Pennisi, "Current-Mode A/D Fuzzy Converter," *IEEE Transactions on Fuzzy Systems*, 10, 4, 533–540, 2002.

3. K. Hirota, "The Concept of Fuzzy Flip-Flop," *IEEE Transactions on Systems, Man and Cybernetics*, 19, 5, 980–997, 1989.

4. K. Hirota and K. Ozawa, "Fuzzy Flip-Flop as a Basis of Fuzzy Memory Modules," *Fuzzy Computing*, M. M. Gupta and T. Yamakawa, Editors, Elsevier Science Publishers, 1998.

5. K. Hirota and K. Ozawa, "Fuzzy Flip-Flop and Fuzzy Registers," *Fuzzy Sets & Systems*, 32, 139–148, 1989.

6. K. Hirota and W. Pedrycz, "Designing Sequential Systems with Fuzzy J-K Flip-Flops," *Fuzzy Sets & Systems*, 39, 261–278, 1991.

7. T. Kettner, C. Heite, and K. Schumacher, "Analog CMOS Realization of Fuzzy Logic Membership Functions," *IEEE Journal of Solid-State Circuits*, 28, 857–861, 1993.

8. B. D. Liu, C. Y. Huang, and H. Y. Wu, "Modular Current-Mode Defuzzification Circuit for Fuzzy Logic Controllers, *Electronics Letters*, 30, 16, 1287–1288, 1994.

9. C. Lu and B. Shi, "Circuit Realization of a Programmable neuron Transfer Function and its Derivative," *Proceedings of the IEEE-INNS-ENNS International Joint Conference on Neural Networks*, IJCNN'00, Como, Italy, July 24–27, 2000, pp.4047–4050.

10. K. Ozawa, K. Hirota, and L. T. Koczy, "Fuzzy Flip-Flop," *Fuzzy Logic Implementation and Applications*, M. J. Patyra and D. M. Mlynek, Editors, John Wiley & Son Ltd., Toronto, 1996, pp. 197–236

11. F. J. Pelayo, I. Rojas, J. Ortega, and A. Prieto, "Current-Mode Analog Defuzzifier, *Electronics Letters*, 29, 9, 743–744, 1993.

12. A. Sanz, "Analog Implementation of Fuzzy Controller," *IEEE International Conference on Fuzzy Systems*, Orlando, FL, USA, 1994, pp. 279–283.

13. Y. Shirai and Yamakawa, "A CAD-Oriented Synthesis of Fuzzy Logic Circuits," *Systems·Computers·Controls*, 15, 5, 76–83, 1984. Translated from Denshi Tsushin Gakkai Ronbunshi, 67-D, 6, 708–714, 1984.

14. K. Tsukano and T. Inoue, "Synthesis of Operationa; Transconductance Amplifier-Based Analog Fuzzy Functional Blocks and Its Application," *IEEE Transactions on Fuzzy Systems*, 3, 1, 61–68, 1995.

15. E. Vittoz, "Present and Future Industrial Applications of Bio-Inspired VLSI Systems," *Proceedings of the 7th International Conference on Microelectronics for Neural, Fuzzy, and Bio-Inspired Systems*, Granada, Spain, April 7–9, 1999, pp. 2–11.

16. T. Yamakawa, "Fuzzy Logic Circuits in Current Mode," *Analysis of Fuzzy Information*, J. C. Bezdek, Editor, 1, 241–261, CRC Press, Boca Raton, Florida, 1987.

17. T. Yamakawa and T. Miki, "The Current Mode Fuzzy Logic Integrated Circuits Fabricated by Standard CMOS Process, *IEEE Transactions on Computers*, c-35, 2, 161–167, 1986.

18. T. Yamakawa, H. Kabuo, "A Programmable Fuzzifier Integrated Circuit-Synthesis, Design, and Fabrication," *Information Sciences*, 45, 75–112, 1988.

19. J. M. Zurada, "Analog Implementation of Neural Networks," IEEE Circuit & Devices, 36–41, September, 1992.

Mixed-Signal Techniques

1. I. Baturone, S. Sánches-Salano, A. Barriga, and J. L. Huertas, "Flexible Fuzzy Controllers Using Mixed-Signal Current-Mode Techniques," *Proceedings of the IEEE International Conference on Fuzzy Systems*, Barcelona, Spain, 1997, pp. 875–880.

2. I. Baturone, S. Sánches-Salano, A. Barriga, and J. L. Huertas, 'Implementation of CMOS Fuzzy Controller as Mixed-signal ICs, *IEEE trasactions on Fuzzy Systems*, 5, 1, 1–19, 1997.

3. I. Baturone, S. Sánches-Salano, A. Barriga, and J. L. Huertas, "Mixed-Signal Design of a Fully Parallel Fuzzy Processor, *Electronics Letters*, 34, 5, 437–438, 1998.

4. S. Bouras, M. Kotronakis, K. Suyama, and Y. Tsividis, "Mixed Analog-Digital Fuzzy Logic Controller with Continuous-Amplitude Fuzzy Inferences and Defuzzification," *IEEE Trasactions on Fuzzy Systems*, 6, 2, 205–215, 1998.

5. E. Franchi, N. Manaresi, R. Rovatti, A. Bellini, and C. Bacarani, "Analog Synthesis of Nonlinear Functions Based on Fuzzy Logic," *IEEE International Journal of Solid-State Circuits*, 33, 6, 885–895, 1998.

Web Resources

Tutorials and Lists

1. Principles of Semiconductor Devices
    http://ece-www.colorado.edu/~bart/book/contents.htm

    An e-book by B. Van Zeghbroeck, Electrical and Computer Engineering Department, University of Colorado at Boulder, USA. Topics covered include:

    - Review of Modern Physics
    - Semiconductor Fundamentals
    - Metal Semiconductor Junctions
    - p-n Junctions
    - Bipolar Transistor
    - MOS Capacitor
    - MOSFETs

2. Operational Transconductance Amplifiers, OTAs
    http://amesp02.tamu.edu/~sanchez/689-Otas-Part1.PDF

    A comprehensive introduction to Operational Transconductance Amplifiers by E. Sánchez-Sinencio, Texas A&M University, Texas, USA. The author introduces transconductance amplifiers and their IC implementation. Detailed circuit analysis and applications are also presented.

3. VLSI Systems: Fuzzy Logic
    http://vlsi.wpi.edu/webcourse/fuzzy/fuzzy.html

    This is Chapter 9, Fuzzy Logic Systems, of the excellent book titled *Design of VLSI Systems* authored by D. Mlynek (Swiss Federal Institute of Technology, Lausanne, Switzerland) and Y. Leblebici (Worcester Polytechnical Institute, Worcester, MA, USA) The complete book is available on the Web. Topics covered in this chapter include:

    - Systems Considerations
    - Fuzzy Logic Based Control Background
    - Integrated Implementations of Fuzzy Logic Circuits
    - Digital Implementations of Fuzzy Logic Circuits
    - Analog Implementations of Fuzzy Logic Circuits
    - Mixed Digital/Analog Implementations of Fuzzy Systems
    - CAD Automation for Fuzzy Logic Circuits Design
    - Neural Networks Implementing Fuzzy Systems

4. Neural Networks Implementation
   www.np.edu.sg/~yck/ANS_L8.pdf

   A tutorial presentation by Yong Chaw Koh, NgeeAnn Polytechnic Singapore. The tutorial explains the term neurocomputing, compares the complexity of neural networks to RAM cells and CPUs, and provides neuron models based on operational amplifiers. It also provides examples of commercial VLSI neural processors and highlights some common applications.

5. Fuzzy Control Methods and their Real System Applications
   www.ic-tech.com/Fuzzy%20Logic/tsld001.htm

   A presentation by IC Tech, Inc. composed of 67 slides, among the topics presented is: How to Select A Special Purpose IC for Fuzzy Control.

6. Computational Neuroscience: Neural circuits in silicon
   www.nature.com/cgi-taf/DynaPage.taf?file=/nature/journal/

   v405/n6789/full/405891a0_r.html

   http://hebb.mit.edu/people/seung/papers/

   news%20and%20views%20nature%20405.pdf

   A two-page document authored by C. Diorioi and R. Rao, appeared in *Nature,* 405, 891–892, 2000. It provides a simple discussion of the biological origins of artificial neural networks. It provides 12 references, some of which are Web-accessible.

7. Fuzzy and Neural Systems Implementation
   http://polimage.polito.it/~lmr/raccolta.html

   A collection of papers covering the various research activities on Neural Networks and Fuzzy Systems carried out jointly at the Department of Electronics of Politecnico di Torino, the Interdepartmental Mechatronics Laboratory of Politecnico di Torino, the Department of Information Engineering of Universita' di Pisa, and at the PERCRO Lab of Scuola Superiore Sant'Anna di Pisa from 1992 to1998. The categories include:

   - Hybrid Intelligent Control and Applications
   - Function Approximation and Applications
   - Neuro-Fuzzy Algorithms
   - Performance Analysis of Neuro-Fuzzy Algorithms
   - Hardware Implementations and Systems
   - Neuro-Fuzzy VLSI Chips
   - Pulse Stream Systems
   - Handwriting Recognition

8. VLSI Architectures and Systems
   www.madess.cnr.it/pubblicazioni/IMPAGINATO03.pdf

   A section from the Italian National Project MADNESS II, 2002. It is composed of several subsections, some related to Fuzzy logic VLSI implementation, including:

   ■ A CAD environment for fuzzy systems development
   ■ VLSI architectures and devices for integrated neuro-fuzzy controllers
   ■ Fuzzy logic methodologies, devices and applications

9. Neural Networks in Hardware: Architectures, Products and Applications
   www.particle.kth.se/~lindsey/HardwareNNWCourse/home.html

   Lecture notes by Clark S. Lindsey available in zipped and HTML formats, updated August, 2002. Topics include:

   Overview, Why Hardware, NNs Applications, Hardware vs Software, Designer's Dilemma, User's Dilemma, NNs Hardware, Case Studies, and References and Links.

10. Commercially Available Hardware List
    www.emsl.pnl.gov:2080/proj/neuron/fuzzy/systems/commercial.html

    This Web site provides links to sources of commercially available hardware for fuzzy logic and other computational intelligence applications.

11. Neural Chips
    http://neuralnets.web.cern.ch/NeuralNets/nnwInHepHard.html

    The site provides Web links to papers that review neural hardware and other sites related to neural hardware. The categories include:

    Commercially available NN Chips and Systems, NN PC accelerators and other cards, and Non-Commercial or prototype NNW Chips and Systems.

Digital Techniques

1. A Collection of Papers on VLSI Processors
   www.bo.infn.it/fuzzy/papers.html

   A set of papers discussing advances in digital implementation of fuzzy logic for embedded systems applications from the Fuzzy Logic Research Group, Physics Department, University of Bologna, Italy. Topics include:

   - Architecture of a 50 MFIPS Fuzzy Processor and the related 1 um VLSI CMOS digital circuits

   - Design and realization of a 50 MFIPS fuzzy processor in 1.0 um CMOS technology

   - Design of a 50 MFIPS digital fuzzy processor and preliminary results of 1.0 um CMOS VLSI MIN-MAX and defuzzifier circuits

   - Design of a VLSI very high speed reconfigurable digital fuzzy processor

   - Fuzzy Logic Oriented to Active Rule Selector and Membership Function Generator for High Speed Digital Fuzzy Processor

   - Digital Membership Function Generators and No-contribute Rule Eliminator for High Speed Fuzzy Architectures

   - High Speed Digital Fuzzy Processor for High Energy Physics Experiment Triggers

   - Design of a Very High Speed Fuzzy Processor by VHDL Language

   - Very Fast VLSI Fuzzy Processor: 2 inputs 1 output

   - A two input fuzzy chip running at a processing rate of 30 ns realized in 0.35 um CMOS technology

2. Implementing Fuzzy Rule Based Systems on Silicon Chips
   www.neuro.sfc.keio.ac.jp/publications/pdf/fuzzy.pdf

   A paper authored by M.-H. Lim and Y. T. Takefuji, which appeared in *IEEE Expert*, 31–45, February, 1990. The authors presented a general block architecture for realizing the approximate reasoning processor. They also showed that the reasoning system's knowledge base is easily programmable and independent of the fuzzy implication relation interpretation used for inferencing.

3. Hardware/Software Co-Design of RISC Processors
    http://jamaica.ee.pitt.edu/Archives/ProceedingArchives/Date/Date98/papers/
    1998/date98/pdffiles/11c_2.pdf

    A paper by V. Salapura and M. Gschwind, Technische Universität Wien,
    Austria. It appeared in the proceedings of the *Design, Automation & Test in
    Europe Conference*, DATE98, Paris, France, February 23–26, 1998, pp. 875–
    782. The authors explained as a case study the definition and evaluation of
    instruction set extensions for fuzzy processing. The proposed instructions
    were evaluated in software and hardware to reach a balanced view of the cost
    and benefit of each instruction. The instruction set was added to a RISC
    processor core. The core was described in VHDL to enable hardware imple-
    mentation using logic synthesis.

4. Fast Digital Fuzzy Controller for an Active Magnetic Bearing
    www.eek.ee.ethz.ch/publications/publication-10.html

    A paper by W. Liebert, Electrical Engineering and Design Laboratory,
    Department of Information Technology and Electrical Engineering of the
    Swiss Federal Institute of Technology in Zürich, Switzerland. The author
    presented the analysis and design of a controller consisting of PID and fuzzy
    controller components. Both controllers were simulated using MATLAB and
    implemented on board using the TMS320C50 fixed point DSP chip.

5. Intel MCS96 Microcontrollers: Fuzzy Logic Applications
    www.mmx.com/design/mcs96/designex/2363.htm

    An application report from Intel outlining the design of a fuzzy logic control-
    ler using the MCS96 microcontroller.

6. Microprocessor Implementation of Fuzzy Logic and Neural Networks
    http://nn.uidaho.edu/pap/2001/Microp_impl_IJCNN01.PDF

    A paper by J. Binfet and B. Wilamowski appeared in the *Proceedings of the
    International Joint Conference on neural Networks*, Washington DC, July
    15–19, 2001, pp. 234–239. The authors present systems that were imple-
    mented on the Motorola 68HC711E9.

7. Fuzzy Logic Control with the Intel 8XC196 Embedded Microcontroller
    www.intel.com/design/mcs96/papers/esc_196.htm

    An 11-page document in PDF format authored by N. Govind of Intel Corpo-
    ration. It discusses the development of a fuzzy inference unit and algorithms
    for fuzzification, rule evaluation and defuzzification of a fuzzy control
    system. Methods to generate optimized fuzzy-based real-time code in
    assembly and C are shown for the Intel 8XC196 microcontroller along with
    discussion of performance and features for fuzzy-based control.

8. A Practical Implementation of a Fuzzy Logic Controller with Motorola 68HC11
   www.elia.ro/susnea/fuzzy.html

   An HTML document authored by Ioan Susnea. The author presents a simple implementation of a fuzzy temperature controller, built around a single chip MC68HC11KA2 microcontroller. It is meant to be a resource for students and designers.

9. Demonstration Model of *fuzzy*TECH Implementation on Motorola 68HC12 MCU
   www.fuzzytech.com/e/e_a_mot.html

   A report by Philip Drake and Jim Sibigtroth of Motorola, and Constantin von Altrock and Ralph Konigbauer of Inform Software Corp. Inform Software Corp. and Motorola created the *fuzzy*TECH MCU-HC12 Edition of Motorola MCU-HC12 which supports both the HC12's fuzzy logic instruction set and the HC12 background debug mode. The authors demonstrate both the usage of the fuzzy logic instruction set and the use of the background debug mode with the *fuzzy*TECH development system. Inform and Motorola have designed an autonomously guided tank as a demonstration model. The authors discuss the fuzzy logic controller design for the tank and the *fuzzy*TECH implementation on the HC12 MCU. More information about *fuzzy*TECH is given in chapter 9.

10. Neuron
    www.castle.uk.co/neuron/models.htm

    This site provides information about embedded computers based on neural networks provided by Castle Technology Ltd, Suffolk, UK. Systems presented include:

    - Neuron 100 – ARM7500 embedded computer built around ARM's 64-MHz 7500FE processor. It provides an ARM 7 central core and supports a wide range of industry standard storage and input/output computer peripherals.

    - Neuron 200 – A StrongARM embedded computer built around the 233-MHz StrongARM processor.

    - Cortex is an optional connector board for Neuron computers. It connects to a Neuron embedded computer to provide connectivity to other devices and can be fitted with a wide range of application-specific connectors and extra circuitry.

11. Neuro-Fuzzy Control and DSPs
    http://polimage.polito.it/~marcello/articoli/microneuro.97.daniela.pdf

    A paper by B. Bona et al., Politecnico di Torino, Italy. It appeared in the *Proceedings of the International Conference on Microelectronics for Neural Networks and Fuzzy Systems*, MICRONEURO'97, Dresden, September, 1997, pp. 113–120. The authors presented a hybrid control system that uses neural and fuzzy techniques and DSP techniques. The TMS320C31 DSP was used for hardware implementation.

12. Fuzzy Logic and PLCs
    www.industrialtext.com/eBooks/Fuzzy_Logic.PDF

    An 18-page e-book from the Industrial Text & Video Company. Topics covered include: Introduction to Fuzzy Logic, Evolution of Fuzzy Logic Control, Fuzzy Logic Principles, Fuzzy Logic Application Example, and Guidelines for Controlling a Complex System.

13. Silicon Axon
    http://citeseer.nj.nec.com/163659.html

    An article authored by B. A. Minch. P. Hasler, and C. Mead. The authors present an axon intended as a building block for pulse-based neural computations involving temporal and spatial correlations of pulses.

14. A Neuro-Fuzzy Real Time Controller for Low Power Embedded Systems
    http://polimage.polito.it/~marcello/articoli/microneuro.94.cintia.pdf

    A paper by L.M. Reyneri et al., which appeared in the *Proceedings of the International Conference on Microelectronics for Neural Networks and Fuzzy Systems*, MICRONEURO'94, Torino, Italy, September 1994, pp. 392–404. The authors describe a neuro-fuzzy real-time controller based on Pulse Stream techniques. The system discussed is a hybrid of two approaches: Neuro-Fuzzy Controllers (implemented using a custom neural chip) and Finite State Automata (implemented using 68Hc11 microcontroller).

15. A Neural Network chip Using CPWM Modulation
    http://polimage.polito.it/~marcello/articoli/iwann.93.pdf

    A paper by M. Chiaberge et al., appeared in the *Proceedings of the International Workshop on Artificial Neural Networks*, IWANN'93, Sitges (E), June 1993, pp. 420–425. The authors described the development and testing of s chip that implemented a neural network using the Coherent Pulse Width Modulation, CPWM, technique.

16. A Comparison between Analog and Pulse Stream VLSI Hardware for Neural Networks and Fuzzy Systems
    http://polimage.polito.it/~marcello/articoli/microneuro.94.comp.pdf

    A Paper by L. M. Reyneri et al. This appeared in the *Proceedings of the International Conference on Microelectronics for Neural Networks and Fuzzy Systems*, Torino, Italy, September 1994, pp. 77–86. The authors used Coherent Pulse Width Modulation, CPWM, as a computing technique to implement neural systems. They compared its performance with systems implemented using analog techniques. Both systems were designed using the same 0.1 µm full custom, double metal, single poly CMOS technology. The authors concluded that in general the two techniques have comparable performance. They also pointed out that the design of CWPM is simpler and multiplexing of CPWM systems is easier.

Analog Techniques

1. Electronic Models of Fuzzy Gates
   http://148.202.12.1/somi/REVISTA/Vol_III/No5/artic8.pdf

   A paper authored by J. L. Pérez and M. A. Banuelos which appeared in the *Journal of Instrumentation and Development* , October 2000, pp. 47–49. In this paper, circuits using op-amps were suggested to implement the functions of fuzzy AND, OR, and NOT. A suggestion for a fuzzy neuron is presented as well.

2. Current Mode CMOS Implementation
   http://nn.uidaho.edu/pap/1995/wcnn95_ota.pdf

   A paper by Y. Ota and B. Wilamowski which appeared in the *Proceedings of World Congress of Neural Networks*, Washington DC, USA, July 17–21, 1995, pp. 480–483. The authors presented a programmable fuzzy system built using current-mode CMOS circuits that implement a Gaussian-type membership function, min-max operators, and a defuzzifier circuit. The membership function is represented by the relation between an input voltage and the output current. The appearance of the Gaussian membership function and position can be controlled by adjusting a reference voltage.

3. Analog Fuzzy Inference Processor
   http://members.fortunecity.com/scienziatopazzo/fuzzy.htm

   A paper by C.T.P. Song, S.F. Quigley, and S. Pammu, published at the *IEEE International Symposium for Circuits and Systems,* Monterey, California, USA, May 31–June 3, 1998, pp. 247–250. The authors presented a design for an analog fuzzy inference processor. It was implemented using CMOS technology.

4. Analog Processing Based on Fuzzy Algorithms
   www.cinstrum.unam.mx/revista/pdfv4n5/art2.pdf

   A paper by J. Castillo et al. This appeared in the *Journal of the Mexican Society of Instrumentation*, 4, 5, 12–18, 2002. The authors presented a fuzzy logic controller circuit on analog CMOS technology. They presented the design and simulation of circuits for membership functions, fuzzy AND, fuzzy OR, and defuzzification.

5. VLSI Implementation of Neural Networks
   http://nn.uidaho.edu/pap/2001/VLSIimpl.pdf

   A paper authored by B. M. Wilamowski, J. Binfet, and M. O. Kaynak, which appeared in

   International Journal of Neural Systems, 3, 10, 191–197, 2000. The authors presented the design and simulation results of two CMOS circuits that implement neural networks. MATLAB was used to develop the training algorithms. The circuits presented were two-input, but the authors indicated that they could be generalized to multidimensional circuits.

6. VLSI Implementation of Fuzzy ART
   http://bach.ece.jhu.edu/pub/papers/aicsp01.pdf

   A paper authored by J. Lubin and G. Cauwenberghs, which appeared in *Analog Integrated Circuits and Signal Processing*, 23, 1–10, 2001. The authors presented a mixed-mode VLSI chip performing unsupervised clustering and classification, implementing models of fuzzy ART.

7. Analog Neuro-Fuzzy Networks for System Modeling and Control
   www.micro.ea.unian.it/Staff/sim/PAPERS/1997ISIS_%20fuzzy.pdf

   M. Conti, S. Orcioni, C. Turchetti, *Advances in Intelligent Systems*, F. C. Morabito (Ed.), IOS Press, Netherlands, 1997. The authors discussed a new type of neural network based on the so-called approximate identities functions. A fuzzy-neural architecture based on this neural network type is then introduced. The authors explained further the reasons for the suitability of the suggested architectures for analog CMOS implementation, along with its SPICE simulations. The authors presented the circuit layout and more details in:

   "A Current-Mode Neuro-Fuzzy Network," *IEEE International Conference on Electronics Circuits and Systems ICECS98,* Lisboa, September, 1998. This may be viewed at: www.micro.ea.unian.it/Staff/sim/PAPERS/1999ISCASfuzzy.pdf

   Implementation results and further details were given in:

   "A Current-Mode Circuit for Fuzzy Partition Membership Functions," *IEEE International Symposium on Circuits and Systems* ISCAS99, Orlando, FL, May 30–June 2, 1999. (www.micro.ea.unian.it/Staff/sim/PAPERS/1999ISCASfuzzy.pdf)

   Further implementation details are given in: "CMOS Analog Implementation of Multidimensional Membership Functions of Neuro-Fuzzy Networks," *IEEE European Conference on Circuit Theory and Design* ECCTD99, Italy, August 29–Sept. 2, 1999.

   www.micro.ea.unian.it/Staff/sim/PAPERS/1997ISIS_%20fuzzy.pdf

8. Adaptation, Learning and Storage in Analog VLSI
   http://bach.ece.jhu.edu/gert/papers/asic96/ASIC96.html

   A paper by Gert Cauwenberghs, Department of Electrical and Computer Engineering, Johns Hopkins University, Baltimore, MD, USA. It appeared in the Proceedings of the 9th Annual IEEE International ASIC Conference, Rochester, NY, September 1996, pp. 237–278. The author presented circuit modules for adaptation and memory, gradient descent learning, and stochastic error-descent learning for analog VLSI implementations.

Mixed Signal Techniques

1. On Designing Mixed Signal Programmable Fuzzy Logic Controllers as Embedded Subsystems in Standard CMOS Technologies
www.dice.ucl.ac.be/~verleyse/papers/sbcci01cd.pdf

   A paper by C. Dualibe, P. Jespers, and M. Verleysen which appeared in the *Proceedings of the 14th Symposium on Integrated Circuits and System Design*, Pirenopolis, Brazil, September 10–15, 2001, pp. 194–200. The author presented a digitally-programmable analog fuzzy logic controller. It is intended for low-power embedded subsystems for applications with band-widths below 8 MHz.

2. A Mixed Signal Neuro-Fuzzy Processor for Embedded Applications
www.micro.ea.unian.it/Staff/sim/PAPERS/1999CSCCfuzzy.pdf

   A paper by M. Conti, S. Orcioni, and C. Turchetti, appeared in the *Proceedings of the IEEE Circuits, Systems, Communications and Computers*, CSCC '99, July 4–8, 1999, Athens. The authors presented a neuro-fuzzy controller implemented using 0.8 µm CMOS analog technology. The architecture is based on an analog device implementing fuzzy rules and a digital interface that allows the device to work as an analog peripheral of a microcontroller.

3. CMOS Design of Adaptive Fuzzy ASICs Using Mixed-Signal Circuits
www.el.uma.es/RafaelNavas/pub_PDF/iscas96.pdf

   F. Vidal-Verdú, R. Navas, and A. Rodríhuez-Vázquez which appeared in the *Proceedings of the IEEE International Symposium on Circuits and Systems*, Atlanta, USA, 1996, pp. 430–433. The authors present a methodology and circuit blocks to realize fuzzy controllers in the form of analog CMOS chips. The authors also provide measurements from a prototype of a fuzzy control-ler chip.

4. Neuro-Fuzzy Architecture for CMOS Implementation
http://nn.uidaho.edu/pap/1999/TIE_neurofuzzy2.pdf

   A paper by B. M. Wilamowski, R. C. Jaeger and M. O. Okyay which ap-peared in the *IEEE Transactions on Industrial Electronics*, 46, 6, 1132–1136, 1999. A non-conventional structure for a fuzzy controller was proposed. It combines fuzzification, MIN operators, normalization, and weighted sum blocks. The structure was implemented as a VLSI chip using n-well CMOS technology.

5. A 5.26 Mflips Programmable Analogue Fuzzy Logic Controller
www.dice.ucl.ac.be/~verleyse/papers/iscas00cd.pdf

A paper by C. Dualibe, P. Jespers, and M. Verleysen which appeared in the *Proceedings of the IEEE Symposium on Circuits and Systems*, Geneva, Switzerland, May 28–31, 2000, 377–380. The authors presented circuit design and simulation results of a digitally-programmable analog fuzzy logic controller. A nine-rule, two-input, and one-output prototype was fabricated and characterized.

6. On-Chip Fuzzy Temperature Control
www.ims.fhg.de/datenblaetter/fuzzy_systems/fuztem/fuztem_e.pdf

A two-page HTML document from Fraunhofer Institut Mikroelektronische Schaltungen und Systeme. It outlines the characteristics of an on-chip fuzzy temperature control that can be used in adaptive fuzzy microsystems.

7. A Modular Programmable CMOS Analog Fuzzy Controller Chip
www.el.uma.es/RafaelNavas/pub_PDF/tcas99.pdf

A paper by A. Rodríguez-Vázquez et al. This appeared in the *IEEE Transactions on Circuits and systems-II: Analog and Digital Signal Processing*, 46, 3, 251–265, 1999. The authors presented a modular fuzzy inference analog MOS chip architecture with on-chip digital programmability. The prototype presented has 16 rules and an operation speed of 2.5 MFLIPS with power consumption of 8.6 mW. It was fabricated using 1-μm CMOS technology. The authors compared the characteristics of the reported prototype with those reported in:

- N. Manaresi et al., "A Silicon Compiler of Analog Fuzzy Controllers: From Behavioral Specifications to Layout," *IEEE Transactions on Fuzzy Systems*, 4, 418–428, 1996, and

- S. Guo et al., "Design and Applications of an Analog Fuzzy Logic Controller," *IEEE Transactions on Fuzzy Systems*, 4, 429–438, 1996.

8. On Designing Mixed-Signal Programmable Fuzzy Logic Controllers as Embedded Subsystems in Standard CMOS Technologies
www.dice.ucl.ac.be/~verleyse/papers/sbcci01cd.pdf

A paper by C. Dualibe et al. This appeared in the Proceedings of the 14th Symposium on Integrated Circuits and System Design, Pirenopolis, Brazil, 10–15 September, 2001, pp. 194–200. The authors presented a prototype of a digitally-programmable analog fuzzy logic controller with a nine-rule, two-input, and one output design. It was implemented using 2.4 μm CMOS Technology with speed up to 5.26 MFLIP.

Software Tools

9.1 Introduction

In this chapter, a selection of software tools for fuzzy systems, neural networks, and neuro-fuzzy systems is presented. The selection can be loosely divided into the following categories: executable files with graphical user interface, executable files that have no graphical user interface, and source codes in C, C++, Java, MATLAB files, etc. The software can also be divided into commercial, shareware, and public domain. They can be also categorized as educational or productivity tools.

The objectives of the software presented here differ. The tools meant for educational purposes illustrate the basic concepts of fuzzy logic and neural networks and help in the visualization of the operation of simple systems. The ones for productivity are meant for system design and analysis. They are all included here since they add educational value and enable the reader to experiment and make judgments based on first-hand experience with a large number of software tools with minimum investment of time and money.

The selection criteria of the software listed can be summarized as follows:

- Relevance to fuzzy systems, neural networks, or both.

- Availability of an English version.

- Documentation is available, or the software can be used intuitively, or both.

- The software is either free or a demo of educational value is available.

- The software and documentation are available through the Web.

The tools are arranged alphabetically for ease of reference, but it would be useful for the reader to examine them starting with the ones that demonstrate general concepts followed by simple productivity tools such as FuzzGen. Fuzzy and neural source codes do, of course, require some knowledge of MATLAB, C, Java, Perl, etc. Commercial productivity tools, such as *fuzzy*TECH, FIDE and Neuro Solutions, can be useful both for education and the design of large-scale projects.

It goes without saying that the software selected here is believed to be reliable and of value. However, inclusion here does not constitute an endorsement or a guarantee for its fitness for a particular purpose.

9.2 Software Overview

1. C and C++ Source Codes

The ART Gallery

cns-web.bu.edu/pub/laliden/WWW/nnet.frame.html

The ART Gallery is a free set of procedures developed by Lars Lidén to be used within other code for implementing many of the ART style neural networks. It supports C procedure calls on both the Unix and DOS platforms, as well as compilation as a dynamic linked library for Windows environments such as Visual Basic.

The Binary Hopfield Net

www.geocities.com/ResearchTriangle/Facility/7620/

The site provides a Java applet developed by Matt Hill. It demonstrates the works of a Hopfield network and the source code is provided as public domain software.

One may use the mouse to enter a pattern by clicking squares inside the rectangle shown. The network stores the pattern; up to approximately 33 random patterns can be stored. After storing some patterns you can then try entering a new pattern (one of the stored patterns, but slightly corrupted) to be recognized, and then watch the network settle.

Neural Networks at your Fingertips

www.geocities.com/CapeCanaveral/1624/

The site provides software simulators developed by Karsten Kutza for eight of the most popular neural network architectures, coded in portable, self-contained ANSI C. The software available for free downloading include:

ADALINE
Simulation of the Adaline network, which is a single-layer backpropagation network. It is trained on a pattern recognition task, to classify a bitmap representation of the digits 0-9 into the corresponding classes. The network recognizes only the exact training patterns. When the application is ported into the multi-layer backpropagation network, fault-tolerance can be achieved. Figure 9.1 shows an example of the patterns to be recognized.

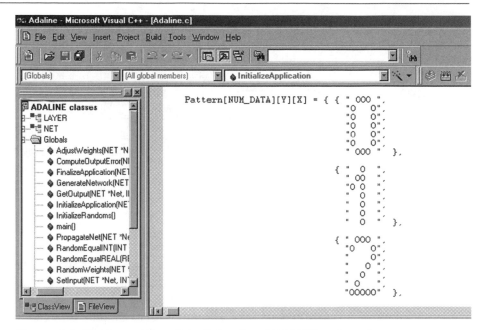

Figure 9.1: Patterns from 0 to 9 for the ADALINE network.

BPN (Backpropagation Network)

This is an implementation of a multi-layer backpropagation network with bias terms and momentum. The network is used to detect structure in time-series. It is presented to the network using a simple tapped delay-line memory. The program can predict future sunspot activity by learning from historical data collected over the past three centuries.

HOPFIELD

This program is based on the Hopfield model. It is used as an auto-associative memory to store and recall a set of bitmap images. Images are stored by calculating a corresponding weight matrix. Then, on presentation of a corrupted image, the network settles on a stored image which is nearest to the input image in terms of Hamming distance.

BAM (Bidirectional Associative Memory)

The bidirectional associative memory allows for a heteroassociative memory to be implemented. It can be viewed as a generalization of the Hopfield model. The software gives, as an example, the association between names and corresponding phone numbers. After training, the network is able to recall the corresponding phone number when presented with a name and vice versa. A degree of fault-tolerance in case of corrupted input patterns is allowed.

BOLTZMAN (Boltzmann Machine)
This is a software implementation of the Boltzmann Machine network. It is used to solve an optimization problem. The weight matrix is chosen such that the global minimum of the energy function corresponds to a solution of a particular instance of the traveling salesman problem.

CPN (Counterpropagation Network)
The counterpropagation network is a competitive network, designed to function as a self-programming lookup table with the additional ability to interpolate between entries. The application is to determine the angular rotation of a rocket-shaped object, images of which are presented to the network as a bitmap pattern. The performance of the network is a little limited due to the low resolution of the bitmap.

SOM (Self-Organizing Map)
In this program the network learns to balance a pole by applying forces at the base of the pole. The behavior of the pole is simulated by numerically integrating the differential equations for its law of motion using Euler's method. The task of the network is to establish a mapping between the state variables of the pole and the optimal force to keep it balanced.

ART1
This program demonstrates the basic feature of the adaptive resonance theory network, which is its ability to adapt when presented with new input patterns while remaining stable at previously learned patterns.

Pittnet
www.pitt.edu/~aesmith/postscript/guide.pdf

Pittnet is a neural network program designed for educational purposes by Brian Carnahan and Alice Smith of the University of Pittsburgh. The compiled C++ source code, the uncompiled C++ source code, and the User's Guide are available for downloading. The software has the following neural network paradigms: backpropagation, Kohonen self-organizing, ART1, and the radial basis function, RBF. The input and output are files of Pittnet ASCII text files.

UNFuzzy
http://ohm.ingsala.unal.edu.co/ogduarte/Software.htm

A program developed by Oscar Duarte, Universidad National de Colombia. It is available in several languages, including English and Spanish.

MS-DOS NN freeware
www.simtel.net/pub/msdos/neurlnet/

This site provides description and links for free downloads of a large number of programs for neural networks, which run under MS-DOS. Some of the programs include:

- ann110; a multi-option NN training program
- bam; a Bidirectional Associative Memory simulation
- bpnn132; a backpropagation neural net with adaptive learning rate
- bps100; a backpropagation simulator
- brain12; a backpropagation neural net simulator
- et; a perceptron simulator
- nerves; Nervous System Construction Kit with C++ source code
- neurfuzz; a program to train NN and fuzzy-engines and generate C source code
- nlmos; an educational simulation for perceptrons, backpropagation, and self-organizing maps
- nnutl101; a C-source code library and tutorial for neural networks.

StarFLIP++
www.dbai.tuwien.ac.at/proj/StarFLIP/

A reusable iterative optimization family of C++ libraries for combinatorial problems with fuzzy constraints designed by the Database and Expert Systems Group, Institute of Information Systems, Vienna University of Technology, Austria. StarFLIP++ is progressing to include the following layered sub-libraries:

- the fuzzy logic inference processor library, FLIP++
- the fuzzy constraint library, ConFLIP++
- the dynamic constraint generation library, DynaFLIP++
- the domain knowledge representation library, DomFLIP++
- several heuristic optimizing libraries, OptiFLIP++
- the user interface, InterFLIP++
- the HTML documentation, DocuFLIP++
- the knowledge-change consistency checker library, CheckFLIP++
- the version-control and test environment for the complete library set, TestFLIP++
- the simulation toolkit, SimFLIP++
- the reactive optimizer, ReaFLIP++
- the neural network extension that allows automatic tuning of fuzzy membership functions, NeuroFLIP++ .

2. A Collection of Fuzzy Logic Based Tools for Analog Circuits Design
www.gte.us.es/usr/chavez/collect.pdf

A paper authored by A. Chávez and L. G. Franquelo appeared in *IEEE Micro*, 60–68, August 1996. It presents a collection of fuzzy logic-based tools in the three main phases of the analog design: topology selection, modeling and optimization, and testing.

3. EazyNN
www.easynn.com/easynnbase.html

EasyNN is a shareware program provided by Stephen Wolstenholme. It can generate multi-layer neural networks from text files or grids. The networks can then be trained, validated and queried. Network diagrams, graphs, input/output data and all the network details can be displayed and printed. Neurons (nodes) can be added or deleted while the network is learning. The software has a step-by-step tutorial, example files, and a help file. A screen shot from the software is shown in Figure 9.2.

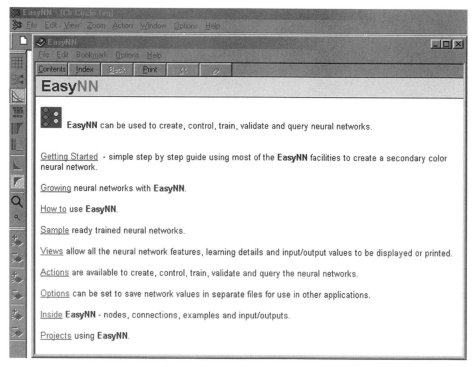

Figure 9.2: EasyNN Help file.

4. FCM, Fuzzy Control Manager
www.transfertech.de/www/soft_e.htm

This is a Windows program that can be used intuitively to define fuzzy systems of any complexity graphically. All relevant data can be displayed while developing, debugging, or optimizing the system. In constructing the system, there is no limit on the number of variables or rules to be defined. The membership functions can be defined by an unlimited number of inflection points (or Singletons for outputs, if desired). Codes for implementing the constructed fuzzy controller can be generated in C-language, assembler, or binary, depending on the FCM version used. FCM versions include:

- FCM-3000 which operates on an FP-3000 adapter board and generates special C code which performs an appropriate access to the FP-3000 processor
- FCM-SOFT which supports the online operation of a VMEbus board containing an FP-3000 processor
- FCM-TEAM which supports a microcontroller board with many analog and digital I/O lines
- FCM-SPS which is used for direct programming of a fuzzy module of a programmable controller containing an FP-3000
- The FCM-Modicon which supports fuzzy programming of an AEG-Modicon FCM-I3HD-CINT which generates codes for the NEC 78K0,78K3, V series microcontrollers
- FCM-I3HD-75X which generates code for microcontrollers of the NEC 75X family
- FCM-I3HD-17K which generates code for microcontrollers of the NEC 17K family
- FCM-PX-1500 which generates code for the Panasonic MN1500 microcontroller family
- FCM-HC11 which generates code for 68HC11 microcontrollers
- FCM-8051 which generates code for microcontrollers of the 8051 family.

A demo with example files is available for downloading and screen shots are available on the Web.

5. FIDE, Fuzzy Inference Development Environment
www.aptronix.com/fide/

FIDE is a development environment for fuzzy logic systems from Aptronix, Inc., Santa Clara, USA. The software can automatically generate fuzzy algorithms in Java, ANSI C, MATLAB M-file, and assembly code for a variety of microcontrollers. Chips supported include: Motorola: 68HC05, 6805, 68HC08, 68HC11, 68HC12, 68HC33x, Intel 80C196 and 80C296 architectures, Siemens: SAE81C99A, and Omron Electronics: FP-3000, FP-5000. A free demo of FIDE and example files are available for downloading. The demo allows the creation and simulation of fuzzy logic models and it helps to learn about the fuzzy inference process. The demo is useful for educational purposes and as a preview of the full version. It is, of course, not meant for use in a production environment; the compiler and code generator are disabled.

6. FLDE, Fuzzy Logic Development Environment
www.flde.com/flde/index.htm

FLDE is a software tool for developing complex embedded control applications. It produces C-code for various target hardware architectures from natural language specifications of the behavior of the controller. Two examples are provided.

The first is Building Embedded Automotive Applications with Fuzzy Logic: A NiCd Battery Charger, by Stylianos S. Sbyrakis, Syndesis Ltd, Athens, Greece. The second is Inverted Pendulum Fuzzy Control.

7. Fuzzy CLIPS
http://ai.iit.nrc.ca/IR_public/fuzzy/fuzzyClips/fuzzyCLIPSIndex.html

FuzzyCLIPS is an extension of the CLIPS (C-Language Integrated Production System) expert system shell from NASA. It was developed by the Integrated Reasoning Group of the Institute for Information Technology of the National Research Council of Canada.

FuzzyCLIPS can deal with exact, fuzzy, and combined reasoning, allowing fuzzy and normal terms to be freely mixed in the rules and facts of an expert system. For non-commercial use, it is available for downloading free of charge, but a licence is required for commercial use.

8. Fuzzy Control Demo
www.csee.wvu.edu/~esazonov/loadswayindex.htm

An interactive example provided by Eduard Sazonov, West Virginia University, USA. It illustrates the use of fuzzy control of a loading crane. In this demonstration, the fuzzy rules controlling the crane can be changed at any time. Three linguistic variables have been defined: angle, distance and power. The linguistic variables and membership functions cannot be changed interactively. The author of the demo points out that unlike the *Fuzzy*Tech example, which works only in one direction of motion, or the Omron example that utilizes complicated anti-sway rules, the given model employs two very simple rules that compensate load sway at the distances close to the destination. For each swing of the load, the crane's trolley is moved in the swing direction to reduce the amplitude. The system's behavior without the compensation for the load sway can be observed by removing the last two rules in the set. A screenshot from the demo is shown in Figure 9.3.

Figure 9.3: Prevention of load sway by a fuzzy controller.

9. FuzzyCOPE
http://kel.otago.ac.nz/software/FuzzyCOPE3/

FuzzyCOPE is a software provided by Knowledge Engineering Laboratory, KEL, University of Otago, New Zealand. It is a hybrid connectionist software environment. It has a graphical user interface to the functionality of the system. The Dynamic Link Libraries that form the core of the system may be used to develop Windows based intelligent systems through the use of the provided programming libraries. Command line tools allow for the creation, training and validation of connectionist structures from the DOS prompt. The software is available for free download. Figure 9.4 shows the contents of the help file in order to get an idea about the potential of the software.

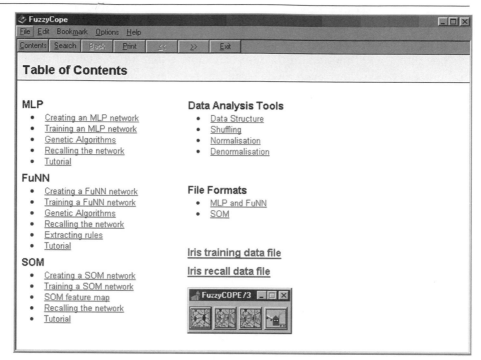

Figure 9.4: The components of the help file of FuzzyCOPE.

10. Fuzzy Decision Tree, FID

www.cs.umsl.edu/~janikow/fid/

FID is a program developed by Cezary Janikow, Mathematics and Computer Science, UMSL St. Louis, MO, USA. It generates a fuzzy logic based decision tree from fuzzy data. The tree can then be used to classify data with unknown classification using several different methods of inference. The ideas used were partially described in a paper by the developer: "Fuzzy Decision Trees: Issues and Methods," *IEEE Transactions on Systems, Man, and Cybernetics*, Vol. 28, Issue 1, pp 1–14, 1998.

11. FuzzGen

www.programmersheaven.com/zone22/cat167/1244.htm

This is a freeware program, and it is probably the earliest fuzzy logic code generator. It allows the user to put together full fuzzy decision-making algorithms graphically, then allows the user to generate code in Pascal, Basic, or C/C++ that will implement the decision process. The software could be used intuitively, but it also has a help file. Figures 9.5 to 9.9 give illustrations from the program to show its capabilities and ease of use.

Figure 9.5: FuzzGen opening screen.

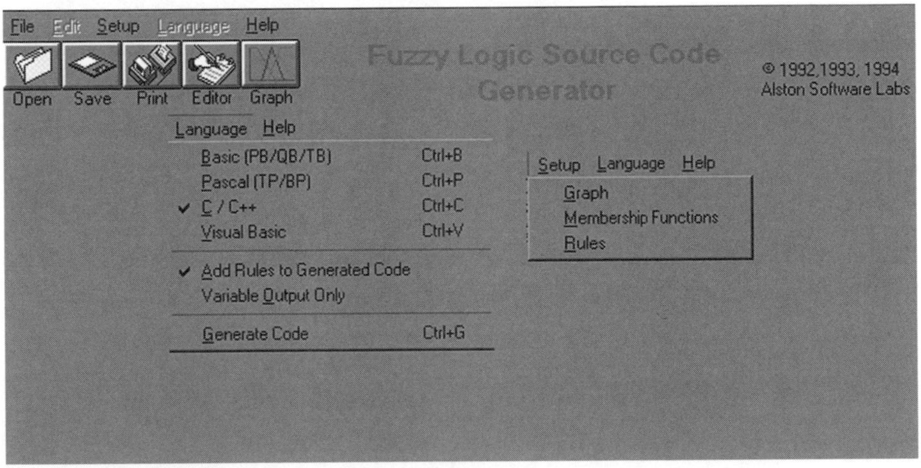

Figure 9.6: Some of the FuzzGen menus.

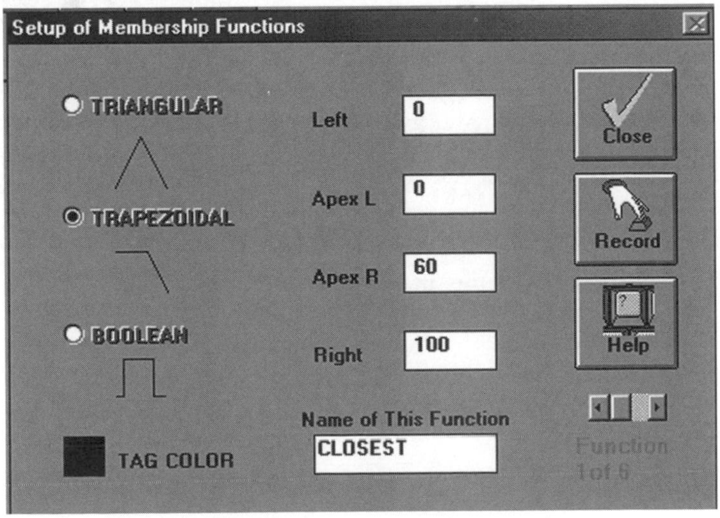

Figure 9.7: Membership selection in FuzzGen.

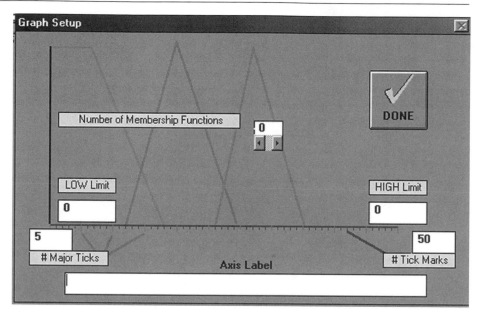

Figure 9.8: Graphical definition of membership functions in FuzzGen.

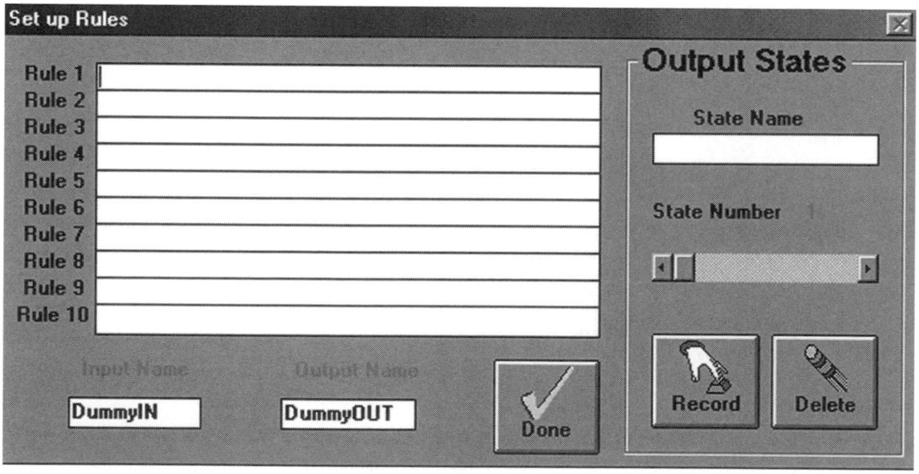

Figure 9.9: Editing fuzzy rules in FuzzGen.

12. Fuzzy Logic Control in VHDL

www.ee.ualberta.ca/~elliott/ee552/studentAppNotes/2000f/vhdl/fuzzyController/

An article by Steve Dillen and Farrah Rashid presented by Duncan Elliott, University of Alberta, Canada. The article is in HTML format with links to source codes.

13. *fuzzy*TECH

www.fuzzytech.com/

*fuzzy*TECH is a leading family of software development tools for fuzzy logic and and neural-fuzzy systems. The software supports both English and German languages. Moreover, the documentation is available in the English, German, and Mandarin Chinese languages. *fuzzy*TECH families of particular interest here include:

- *fuzzy*TECH Editions for General Target Hardware including *fuzzy*TECH Professional and *fuzzy*TECH online editions.
- *fuzzy*TECH MCU Editions for Embedded Control including:
  - *fuzzy*TECH MCU-HC05/08 Edition which supports all Motorola 68HC05xx and 68HC08xx families of microcontrollers
  - *fuzzy*TECH MCU-ST6 Edition which supports all STMicroelectronics ST6 family microcontrollers
  - *fuzzy*TECH MCU-MP Edition which supports all Microchip's microcontrollers
  - *fuzzy*TECH MCU-51 Edition which supports all 8051 and 80251 microcontrollers
  - *fuzzy*TECH MCU-374xx Edition which supports all 8-bit Mitsubishi microcontrollers
  - *fuzzy*TECH MCU-HC11/12 Edition which supports all Motorola 68HC11xx and 68HC12xx families of microcontrollers.
  - *fuzzy*TECH MCU-96 Edition which supports all Intel MCS-96 microcontrollers
  - *fuzzy*TECH MCU-320 Edition which supports Texas Instruments Digital Signal Processors
- *fuzzy*TECH IA Editions for Industrial Automation including *fuzzy*TECH IA-S5/7 Edition for Siemens SIMATIC Programmable Logic Controllers S5.

Application examples are discussed in numerous reports by C. von Altrock including:

- Fuzzy Logic in Automotive Engineering, Circuit Cellar, 88, 1–9, 1997, available at: www.circuitcellar.com/pastissues/articles/misc/88constantin.pdf

- Practical Fuzzy Logic Design, Circuit Cellar, 75, 1–5, 1996, available at: www.circuitcellar.com/pastissues/articles/misc/75constantin1.pdf

- A Fuzzy-Logic Thermostat, Circuit Cellar, 75, 1–5, 1996, available at: www.circuitcellar.com/pastissues/articles/misc/75constantin2.pdf

A *fuzzy*TECH library of technical applications is also available online. It includes reports on topics such as:

- Practical Design, Industrial Automation, Coal Power Plant, Refuse Incineration Plant, Complex Chilling Systems, Water Treatment System, Monitoring Glaucoma, AC Induction Motor, Truck Speed Limiter, Medical Shoe, Fuzzy in Appliances, Automotive Engineering, Antilock Braking System, Aircraft Flight Path, Nuclear Fusion, Motorola 68HC12 MCU, Traffic Control, Fuzzy Logic Standardization, and Sonar Systems, and Fuzzy Logic Design: Methodology, Standards, and Tools.

Further information, discussions and illustration of *fuzzy*TECH applications are available at: www.neurotech.com.sg/html/archive/supply_archive_files/fuzzy.html

A demo of *fuzzy*TECH along with example files is available for downloading. The following are some illustrative screen shots from *fuzzy*TECH.

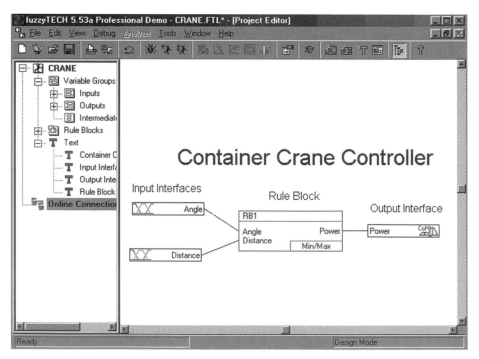

Figure 9.10: Opening one of the *fuzzy*TECH example files.

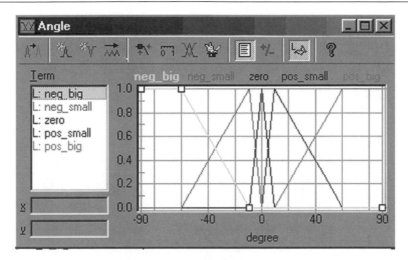

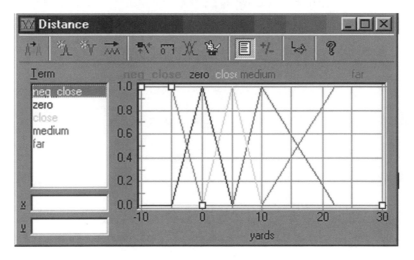

Figure 9.11: Defining inputs for the Container Crane Controller example.

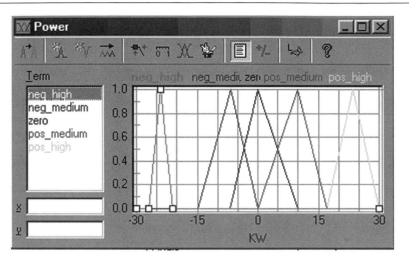

Figure 9.12: Defining the output of the Container Crane Controller example.

#	IF			THEN	
	Angle	Distance		DoS	Power
1	pos_small	zero		1.00	neg_medium
2	zero	zero		1.00	zero
3	pos_small	close		1.00	neg_medium
4	zero	close		1.00	zero
5	neg_small	close		1.00	pos_medium
6	neg_small	medium		1.00	pos_high
7	neg_big	medium		1.00	pos_medium
8	zero	far		1.00	pos_medium
9	neg_small	far		1.00	pos_high
10					
11					
12					
13					
14					
15					
16					
17					
10					

Figure 9.13: *fuzzy*TECH Fuzzy rules editor.

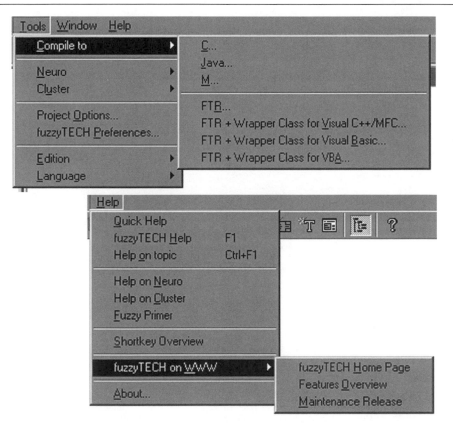

Figure 9.14: Some of the menus of *fuzzy*TECH showing important features of the software.

14. havBpETT
www.hav.com/default.html

havBpETT demonstrates the use of a DLL version of the havBpNet++ C++ neural network class library. The demo allows the user to describe, train, and save a simple feed-forward, recurrent or sequential neural network. The downloadable demos are the same as the fully functional simulator except that the network-save function is disabled. Screen-shots to illustrate the main user interface screens of the havBpETT demo are also available on the Web site.

15. HyperLogic Demos

www.hyperlogic.com/demos.html

The site offers several demo programs, some of which run under Windows and some under MS-DOS. The main software is CubiCalc. A complete working model of that software is available with all project save and data export capabilities disabled. The demo has several example files that can be opened and examined as shown in Figure 9.15. The problems are described in detail and an example of the results of running the simulation of driving in a circle is shown in Figure 9.16.

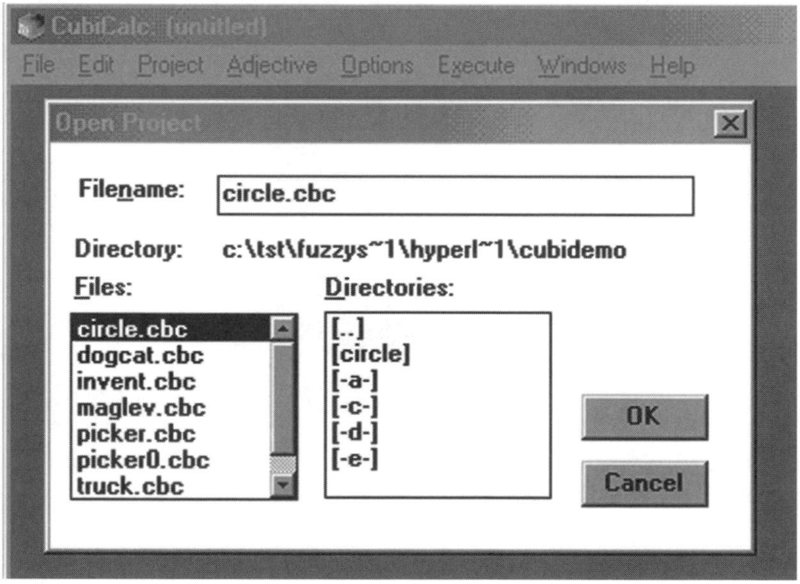

Figure 9.15: Opening one of the examples provided with CubiCalc.

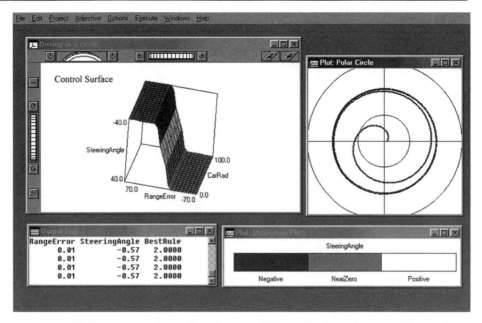

Figure 9.16: Example of results that could be displayed using CubiCalc.

The site also provides an MS-DOS file that explains how the OWL Neural Network Library of C-subroutines works. It includes examples using Backpropagation and Competitive Learning. The source code for a sample program is also provided. In addition to Fuzzy Engineering Examples, numerous MS-DOS and Windows files that illustrate some of the neural networks operations and fuzzy logic applications are provided, including a truck backer-upper.

16. Jan's Fuzzy System, JFL
http://inet.uni2.dk/~jemor/jfs.htm

JFs is a development environment for the programming language JFL, presented by Jan Mortensen, Frederiksberg, Denmark as freeware. The environment has tools to compile, run, improve and convert JFL programs. JFL combines features from traditional programming languages like fuzzy logic and machine learning. A compiled JFs-program can be executed by a command line from C-programs, or from a DLL. It can be converted to an HTML file with the program converted to JavaScript. It can also be converted to a source code file and included in C-programs.

17. Java Related

Joon NN
http://joone.sourceforge.net/

Joone is a free neural net framework to create, train and test neural nets. It is based on Java technologies. Joon NN is composed by a central engine that is the pivot of all existing applications and those that will be developed.

Applets for Neural Networks and Artificial Life
www.aist.go.jp/NIBH/~b0616/Lab/Links.html

A collection of Java applets provided by Akio Utsugi, National Institute of Advanced Industrial Science and Technology, Japan. The applets are classified into several categories, including:

- **Competitive Learning**
  - Vector Quantizer, VQ
  - Gaussian Mixture Model
  - Multinomial Mixture Model
  - Several Self-Organizing Maps, SOM
  - Interactive SOM
  - Pattern Recognition using Kohonen Featuremap
  - Several nets for TSP
  - Neural Competitive Models Demo
  - Comparison between various kinds of competitive learning with topology reformation.
  - Bayesian Self-Organizing Maps

- **Backpropagation Learning**
  - Learning of Function Approximation
  - Interactive Tutorials on Artificial Neural Learning
  - Animated Neural Network Learn 3-D plane
  - Artificial Neural Network Handwriting Recognizer
  - OCHRE - Optical Character Recognition

- **Neural Nets for Constraint Satisfaction and Optimization**
  - Hopfield Net
  - 8-Queen Problem
  - Boltzmann Machine
  - Content Addressing Memory by Hopfield Net
  - Brain Wave
  - Pattern reconstruction
  - and more.

18. LabVIEW Related

LabVIEW is short for Laboratory Virtual Instrument Engineering Workbench. It is a graphical program development application from National Instruments. LabVIEW programs are called *virtual instruments*, VIs, because their appearance and operation mimic actual instruments. A free evaluation version of LabVIEW is available from National Instruments at: www.ni.com/downloads/

Useful LabVIEW resources are available at: www.ni.com/downloads/

A list of all off-the-shelf VIs for a specified purpose is available at: www.mooregoodideas.com/FAVIs/

Free VIs are available from REAL Controls, Inc. at: www.realcontrols.com/download.htm

Commercial fuzzy VIs are available at: www.dataengine.de/english/

A summary report of a PhD thesis titled "Control of a Pneumatic Servosystem Using Fuzzy Logic" by H. M. Llagostera, UPC, Barcelona, Spain,where LabVIEW is used in the fuzzy control scheme is available at: http://fluid.power.net/techbriefs/papers/proc_moreno.pdf

ARGE Simulation News, ARGESIM, is a non-profit working group that provides the infrastructure and administration for dissemination of information on modeling and simulation in Europe. It is located at Vienna University of Technology, Simulation Department. Several comparisons have been defined in Simulation News Europe. Comparison # 9 titled "Fuzzy Control of a Two Tank System" includes the use of LabVIEW. It is available at: http://argesim.tuwien.ac.at/comparisons/

19. Linguistic Fuzzy Logic Controller, LFLC
http://ac030.osu.cz/irafm/lflc/lflc.html

Linguistic Fuzzy Logic Controller is provided by the Institute for Research and Applications of Fuzzy Modeling, University of Ostrava, Czech Republic. The software has two different basic inference methods: one based on the interpretation of the IF-THEN rules as linguistically described logical implications, and the other is the standard max t-norm rule which is the interpolation of an unknown function. A demo of the software is available for download. The demo has example files and a detailed help file.

20. Matlab Related

MATLAB is a commercially available (www.mathworks.com) interactive environment and programming language for scientific and technical computation; it also has a student version at a reduced cost. It allows solution of many numeric problems in much less time than it would have taken to write a program in C or C++ (for example). MATLAB allows users to build their own reusable tools. One can create one's own special functions and programs to run in a MATLAB environment; such programs are referred to as M-files. MATLAB functions created to deal with a certain class of problems could be grouped together, leading to the concept of a toolbox. A toolbox in this context refers to a specialized collection of M-files for working on a particular class of problems. Numerous toolboxes are available commercially, for example:

- COMMUNICATIONS TOOLBOX, for the design and analysis of communication systems

- CONTROL SYSTEM TOOLBOX, for the design and analysis of feedback control systems

- CURVE FITTING TOOLBOX, for model fitting and analysis

- DATA ACQUISITION TOOLBOX, for acquiring and sending out data from plug-in data acquisition boards

- DATABASE TOOLBOX, for exchanging data with relational databases

- DATAFEED TOOLBOX, for acquiring real-time financial data from data service providers

- EXTENDED SYMBOLIC MATH TOOLBOX, for performing computations using symbolic mathematics and variable precision arithmetic

- FILTER DESIGN TOOLBOX, for the design and analysis of advanced floating-point and fixed-point filters

- **FUZZY LOGIC TOOLBOX**, to support the design and analysis of fuzzy logic based systems. It supports all phases of the process, including development, research, design, simulation, and real-time implementation. It uses graphical user interfaces, GUIs, to provide an intuitive environment to guide the user through the steps of fuzzy inference system design. Functions are provided for many fuzzy logic methods, such as fuzzy clustering and adaptive neuro-fuzzy learning. Demo tutorials are provided for: Fuzzy Logic Controller in Simulink, Graphical Editors, Noise Cancellation, and Subtractive Clustering. Tutorial information is provided by Albert Oller i Pujol, Escola Tècnica Superior d'Enginyeria (ETSE) at: www.etse.urv.es/~aoller/fuzzy/fuzzy_logic.htm#4.

A tutorial based on the fuzzy logic toolbox is provided in PDF format by N. H. Koivo, Systeemitekniikan laboratorion kurssit. It is available at: www.control.hut.fi/Kurssit/AS-74.115/Material/fuzzy2.pdf

- IMAGE PROCESSING TOOLBOX, for performing image processing, analysis, and algorithm development

- INSTRUMENT CONTROL TOOLBOX, to control and communicate with test and measurement instruments

- LMI CONTROL TOOLBOX, to design robust controllers using convex optimization techniques

- MAPPING TOOLBOX, to analyze and visualize geographically based information

- MODEL PREDICTIVE CONTROL TOOLBOX, to control large, multivariable processes in the presence of constraints

- MODEL-BASED CALIBRATION TOOLBOX, to calibrate complex power-train systems

- μ-ANALYSIS AND SYNTHESIS TOOLBOX to design multivariable feedback controllers for systems with model uncertainty

- **NEURAL NETWORK TOOLBOX,** to provide tools for the design, implementation, visualization, and simulation of neural networks. It provides comprehensive support for many proven network paradigms, as well as a graphical user interface. Demo tutorials are available for Classification Using a Probabilistic Neural Network, Function Approximation with Radial Basis Networks, Introduction to the Neural Network Toolbox, and Signal Prediction

- OPTIMIZATION TOOLBOX, to solve standard and large-scale optimization problems

- PARTIAL DIFFERENTIAL EQUATION TOOLBOX, to solve and analyze partial differential equations

- ROBUST CONTROL TOOLBOX, to design robust multivariable feedback control systems

- SIGNAL PROCESSING TOOLBOX, to perform signal processing, analysis, and algorithm development

- STATISTICS TOOLBOX, to apply statistical algorithms and probability models

- SYMBOLIC MATH TOOLBOX, to perform computations using symbolic mathematics and variable precision arithmetic

- SYSTEM IDENTIFICATION TOOLBOX, to create linear dynamic models from measured input-output data

- VIRTUAL REALITY TOOLBOX, to animate and visualize Simulink systems in three-dimensions

- WAVELET TOOLBOX, to analyze, compress, and de-noise signals and images using wavelet techniques.

There are even more commercially available toolboxes provided by a third party. In addition, and most importantly, there are toolboxes and M-files developed by top researchers and placed in the public domain. Moreover, an environment similar to MATLAB has also been developed and placed in the public domain. The following sections give a short overview of some of these public domain tools.

GNU Octave
www.octave.org/

GNU Octave is a high-level language similar to MATLAB but available freely. It was written by John W. Eaton with contributions from numerous other individuals listed on the Web site. A GNU Octave Repository is accessible at: http://octave.sourceforge.net/

It is a central location for custom scripts, functions and extensions for GNU Octave.

Useful relevant files are also available from MatLinks/Chorus at: http://sourceforge.net/projects/matlinks

This is an open source project that provides GNU Octave and MATLAB/Simulink toolboxes available in the public domain.

NNSYSID Toolbox
http://kalman.iau.dtu.dk/research/control/nnsysid.html

This is a set of MATLAB tools for neural network based identification of nonlinear dynamic systems. The set is composed of M- and MEX-files for training and evaluation of multilayer perceptrons. There are functions for training of ordinary feedforward networks as well as for identification of nonlinear dynamic systems and for time-series analysis. The toolbox requires MATLAB 5.3 or higher, but it is completely independent of the Neural Network Toolbox and the System Identification Toolbox.

SOM Toolbox for MATLAB
www.cis.hut.fi/projects/somtoolbox/

SOM is a freeware package for the MATLAB environment. Numerous individuals have contributed to the SOM Toolbox. They are or were employed by the Laboratory of Information and Computer Science at the Helsinki University of Technology. The toolbox comes with several demos, including:

To illustrate some of the capabilities of the toolbox, illustrations from parts of the code provided in the first demo and the resulting output are shown in the following.

som_demo1

A self-organized map, SOM, maps the training data. It consists of neurons located on a regular map grid. The lattice of the grid can be either hexagonal or rectangular. The diagram illustrated in Figure 9.17 results from running the following segment in the MATLAB environment:

```
subplot(1,2,1)
som_cplane('hexa',[10 15],'none')
title('Hexagonal SOM grid')

subplot(1,2,2)
som_cplane('rect',[10 15],'none')
title('Rectangular SOM grid')
```

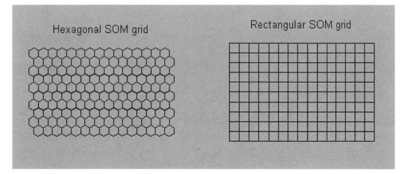

Figure 9.17: The generated SOM grid.

Then, running the segment:

```
subplot(1,3,1)
som_grid(sMap)
axis([0 11 0 11]), view(0,-90), title('Map in output space')

subplot(1,3,2)
plot(D(:,1),D(:,2),'+r'), hold on
som_grid(sMap,'Coord',sMap.codebook)
title('Map in input space')
```

results in the arrangement shown in Figure 9.18.

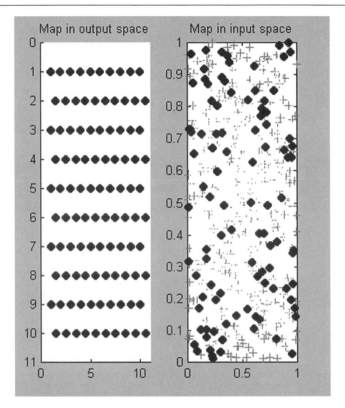

Figure 9.18: The black dots show positions of map units (neurons), and the gray lines show the connections. The map was initialized randomly and hence the positions in the input space are completely disorganized. The crosses are training data.

The training is based on:

- competitive learning, where the prototype vector most similar to a data vector is modified so that it becomes more similar to it, and

- cooperative learning, where not only the most similar prototype vector, but also its neighbors on the map are moved towards the data vector.

This leads to the results shown in Figure 9.19.

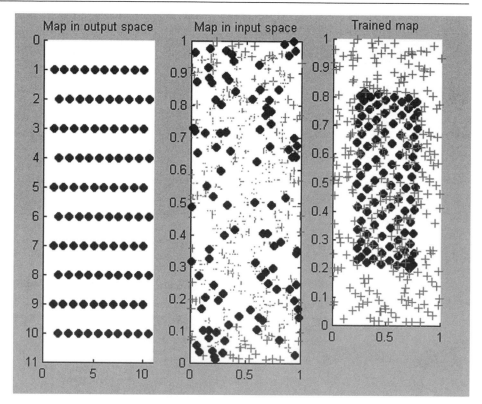

Figure 9.19: The map organizes and folds to the training data.

M-files for Fuzzy Membership Optimization
http://academic.csuohio.edu/simond/fuzzyopt/

This site makes available numerous M-files developed by Dan Simon, Department of Electrical Engineering, Cleveland State University, USA. They demonstrate fuzzy membership function optimization using gradient descent and Kalman filters. The task which is considered by these M-files is an automotive cruise control system. The M-files are downloadable in zip format. In addition, the site provides links to references in the area of optimization of fuzzy logic systems. The files include:

VehicleControl.m, a fuzzy vehicle cruise control program. Running this file will simulate a fuzzy cruise controller for the situation where a vehicle suddenly goes from a road grade of 0% to 10%. One can change any of the details of the simulation (step size, simulation length, vehicle mass, road grade, etc.). Most of the parameters in the file are self-explanatory. Figure 9.20 shows an example of the result of running the file.

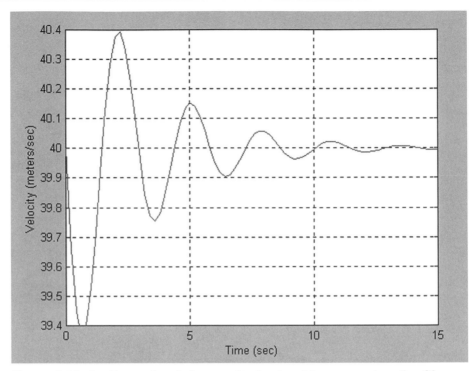

Figure 9.20: An Example of the result obtained from running the file VehicleControl.m.

FuzzCalc.m, a general-purpose fuzzy logic engine that computes defuzzified outputs corresponding to fuzzy inputs. This file can also compute the derivatives of the outputs with respect to the membership function parameters. This routine uses triangular membership functions.

FuzzInit.m, initializes the rule base and membership functions for a fuzzy logic system.

GradeCalc.m, a general-purpose routine that computes the membership grade of a number in a triangular fuzzy set.

PlotMem.m, plots triangular membership functions based on the membership parameters stored in a file.

VehicleGrad.m, optimizes a fuzzy cruise control system using gradient descent.

VehicleKalman.m, optimizes a fuzzy cruise control system using a Kalman filter.

Dmatrix.m, is an auxiliary routine that creates a matrix that is used when the optimization is carried out with some normal constraints.

Engineering Applications of Soft Computing
www.fmt.vein.hu/softcomp/software.html

Numerous M-files provided by the Department of Process Engineering, Folyamatmérnöki Tanszék. They relate to the work presented in:

- J. Abonyi and J.A. Roubos, M. Oosterom, F. Szeifert, "Compact TS-Fuzzy Models through Clustering and OLS plus FIS Model Reduction", *FUZZ-IEEE'01*, Sydney, Australia, 2001.

- J. Abonyi, R. Babuska, F. Szeifert, "Fuzzy Modeling with Multidimensional Membership Functions: Constrained Identification and Control Design", *IEEE Systems, Man and Cybernetics*, Part B, Oct, 2001.

- J. Abonyi and R. Babuska and M. Ayala Botto and F. Szeifert and N. Lajos, "Identification and Control of Nonlinear Systems Using Fuzzy Hammerstein Models", *Industrial and Engineering Chemistry Research*, 39, 4302-4314, 2000.

Type-2 Fuzzy Logic Software
http://sipi.usc.edu/~mendel/software/

This is a freeware collection of M-files provided by Nilesh N. Karnik, Qilian Liang and Jerry M. Mendel, University of Southern California. The collection has four sections: general type-2 fuzzy logic systems, interval type-2 fuzzy logic systems, type-1 fuzzy logic systems, and NEW type-reduction.

NEFCON for MATLAB
http://fuzzy.cs.Uni-Magdeburg.de/nefcon/nef_mat.html

NEFCON is an implementation of a neural fuzzy controller based on a neuro-fuzzy controller using a fuzzy-perceptron. It requires a MATLAB/Simulink environment. NEFCON is able to learn fuzzy sets and rules by a reinforcement learning algorithm. The model was developed by the Fuzzy Systems research group at the Technical University of Braunschweig, Germany. The software is free for non-commercial use. It has online documentation with screen shots.

MATLAB Software Tool for Neuro-Fuzzy Identification and Data Analysis
http://iridia.ulb.ac.be/~gbonte/software/Local/FIS.html

This software was developed by Gianluca Bontempi and Mauro Birattari, ULB Brussels, Belgium. The software trains a fuzzy architecture on the basis of a training set of N (single) output-(multi) input samples. The software is free.

21. Motorola

Motorola Fuzzy logic engine for the 68HC11

www.programmersheaven.com/zone5/cat26/1288.htm

e-book from Motorola

http://faculty.petra.ac.id/resmana/private/fuzzy/

Screen shots from the Motorola e-book are shown in Figures 9.21 and 9.22.

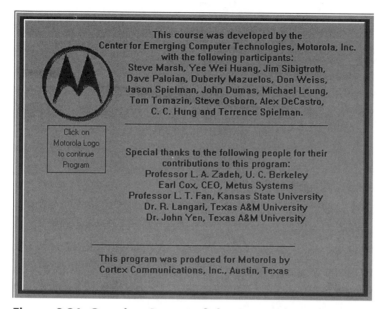

Figure 9.21: Opening Screen of the Motorola Tutorial.

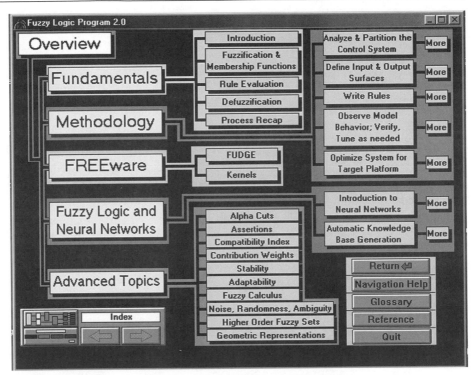

Figure 9.22: Menu selection of the Motorola Tutorial.

22. NEFCLASS, Neurofuzzy Classification

http://fuzzy.cs.uni-magdeburg.de/nefclass/nefclass.html

NEFCLASS is a software provided by Detlef Nauck, Chief Research Scientist and Team Leader in the Intelligent Systems Lab of BTexact Technologies Advanced Research Department, Adastral Park, United Kingdom. It is freely available for scientific and personal use. The software is intended to be used for data analysis by neuro-fuzzy models. It can learn fuzzy rules and fuzzy sets by supervised learning. It can represent a fuzzy classification system, learn fuzzy classification rules incrementally, and learn fuzzy sets by using simple heuristics.

23. Neural Networks and Fuzzy Systems Software
fuzzy.cs.uni-magdeburg.de/software.html

This site provides links to download numerous excellent software programs developed by the research group on Neural Networks and Fuzzy Systems at the Institute of Knowledge Processing and Language Engineering, Faculty of Computer Science, University of Magdeburg, Germany. The programs available relate to: Neuro-Fuzzy Control, Neuro-Fuzzy Function Approximation, Fuzzy Clustering, Neuro-Fuzzy Data Analysis, Multilayer Perceptron Training Visualization, Learning Vector Quantization Visualization, and Self-Organizing Map Training Visualization. For example, one of the available programs is the Multilayer Perceptron Training Visualization, named XMLP. The software helps to visualize the training process of a simple multi-layer perceptron for two logical and two simple real value functions. Figures 9.23 and 9.24 show screen shots from that program.

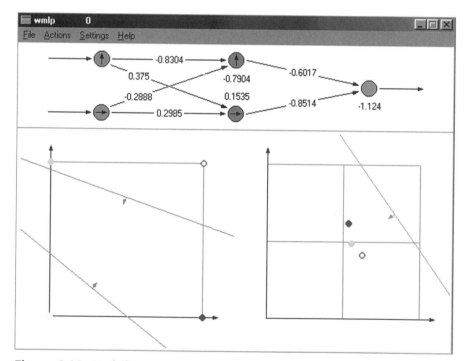

Figure 9.23: Multilayer perceptron training.

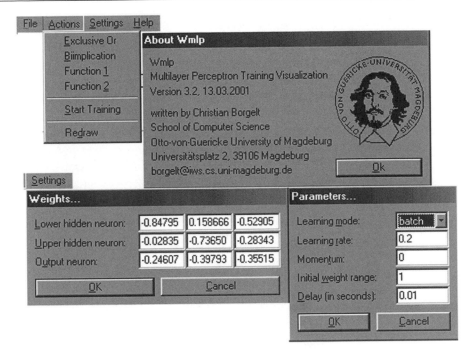

Figure 9.24: Illustration of some menu selections from XMLP.

24. NeuroDesigner

www.neurodesigner.com/

NeuroDesigner is a comprehensive family of Java-based programs for neural network applications. A trial version of NeuroDesigner Editor is available for download. This version is operational for 30 days or 100 times, depending on usage. Moreover, the trial version has limited capabilities such as: the network cannot have more than 10 input and 25 output neurons, or have more than one hidden layer with four hidden neurons. Among the software capabilities highlighted are: embedding neural networks into existing systems and fetching data from various sources.

25. Neuro Solutions
www.nd.com/

NeuroSolutions is a graphical neural network development tool from NeuroDimension in Gainesville, Florida, USA. The Web page has links to a free trial version and a tour of the software complete with screen shots. The following are highlights from the tour. NeuroSolutions enables the user to create a neural network model for a given set of data. It supports numerous network architectures including: CANFIS Network (Fuzzy Logic), Multilayer Perceptron (MLP), Self-Organizing Map Network (SOM), Generalized Feed Forward, Generalized Recurrent Network , Learning Vector Quantization (LVQ), and more. The software has two separate wizards that enable the user to automatically build a neural network:

- **NeuralExpert**, where the design specifications are centered on the type of problem in which the user is interested, such as Classification, Prediction, Function Approximation, or Clustering, and

- **NeuralBuilder**, where the design specifications are centered on the specific neural network architecture the user wishes to construct.

The software allows probes that provide real-time access to all internal network parameters and data, including inputs/outputs, errors, gradients, weights, etc.

It also allows the user to optimize attributes such as the learning rates, weights, number of hidden neurons, etc. The Professional and Developer levels of NeuroSolutions allow the automatic generation of C++ source code for the neural network designed.

26. NICO Artificial Neural Network Toolkit
www.speech.kth.se/NICO/index.html

The NICO (Neural Inference COmputation) toolkit was developed by Nikko Ström at the Department for Speech, Music, and Hearing at KTH, Stockholm, Sweden. It is a general purpose toolkit for constructing artificial neural networks and training with the back-propagation learning algorithm. Units (neurons) are organized in groups and the group is a hierarchical structure; groups can have sub-groups or other objects as members. This makes it easy to specify multi-layer networks with arbitrary connection structure and to build modular networks. There is no support for viewing the networks graphically.

27. Rigel Corporation

www.rigelcorp.com/8051soft.htm

This site provides a variety of 8051 software. The software, help files, examples, and related text files are offered at no charge for non-commercial use. The software available includes Reads51, which is an integrated application software development system. It allows writing, compiling, assembling, debugging, downloading, and running applications software in the MCS-51 language. It contains a C-compiler, relative assembler, linker/locator, editor, chip simulator, assembly language debugger, and host-to-board communications in a menu-driven environment. Information about Fuzzy Logic LabPac is provided, which is a package for fuzzy logic applications using the 8051 family of microcontrollers.

28. Stuttgart Neural Network Simulator (SNNS)

www-ra.informatik.uni-tuebingen.de/SNNS/

Stuttgart Neural Network Simulator, SNNS is a software simulator for neural networks on Unix workstations developed at the Institute for Parallel and Distributed High Performance Systems (IPVR) at the University of Stuttgart, Germany. It consists of two main components: a simulator kernel written in C, and a graphical user interface.

The simulator kernel operates on the internal network data structures of the neural nets and performs all learning and recall operations. It can also be used without the other parts as a C program embedded in custom applications. The network architectures and learning procedures supported include: Backpropagation, Counterpropagation, Generalized Radial Basis Functions, ART1, ART2, ARTMAP, and more.

29. WinNN

www.geocities.com/SiliconValley/Lab/9052/winnn.htm

This software incorporates a user friendly interface with a powerful computational engine. It can implement feedforward multi-layered NN and uses a modified fast back-propagation for training and various neuron functions. It has extensive online help, and various neuron functions. The software has some example files. It is available for a free 60 day trial.

30. Xfuzzy

www.imse.cnm.es/Xfuzzy/

Xfuzzy is an environment for the design, verification, and synthesis of fuzzy-logic-based systems. It can be compiled and executed on Windows using:

- Cygwin (http://sources.redhat.com/cygwin/)
- XFree86 (http://sources.redhat.com/cygwin/xfree/)

Quizzes

Conclude whether the statements in each of the following sections are true or false; correct or explain the false ones.

Embedded Systems

1. Embedded computational systems are restricted to consumer electronics and robotics.

2. Moore's law is the reason that IC technology doubles its performance every 18 months.

3. Embedded systems' design metrics are measurable features of the implemented system.

4. Power consumption is one of the design metrics of embedded systems.

5. Design metrics are typically positively correlated.

6. Embedded systems can only be implemented using either microprocessors or microcontrollers.

7. The physical properties of Gallium Arsenide lead to inherently faster devices than those based on Silicon.

8. BiMOS devices are semiconductor devices that are biologically inspired.

9. Fuzzy logic would lead to *smarter* consumer products.

10. Using fuzzy logic in the design of an embedded system could reduce the time-to-market of the product.

Answers:

False:

1 (there are numerous applications including communications, business, appliances, etc.),
2 (don't confuse the moon with the finger that points to it),
5 (negatively; improving one feature could lead to worsening of another),
6 (other hardware can be used),
8 (based on bipolar and CMOS technologies).

Fuzzy Sets & Relations

1. The membership function is another name for a probability distribution.

2. Membership values in a fuzzy set do not have to add up to one.

3. A classical set would have a Gaussian-type membership function.

4. DeMorhan's theorems apply to both crisp and fuzzy sets.

5. The membership function of a crisp set cannot be defined.

6. The membership function of a fuzzy set has to be of a triangular shape.

7. The law of the excluded middle does not apply to fuzzy sets.

8. A fuzzy set is a collection of items about which nothing is known.

9. All defuzzification algorithms yield the same outcomes.

10. There is no measure of fuzziness.

11. The CON operation makes a fuzzy set less fuzzy.

12. There is no definition of the cardinality of a fuzzy set.

13. The order of operation of Cartesian multiplication is important in classical sets only.

14. In general, the intersection of fuzzy sets results in a crisp set.

15. In general, the union of fuzzy sets results in a fuzzy set.

16. No two fuzzy sets can be equal.

17. A set and its complement have to be identical for a set to be defined as a crisp set.

18. The support of a fuzzy set is a crisp set.

Answers:

False:

1 (fuzziness is not probability)
3 (all elements in the universe of discourse will have a membership of either one or zero)
5 (see above)
6 (trapezoidal, Gaussian and discrete values are common)
8 (precise knowledge is needed to assign the appropriate membership value)
9 (different algorithms will not result in the same numerical values)
10 (there are several possible measures; one possibility relates to how far the deviation is from the excluded middle)
12 (there is more than one possible definition)
13 (it is important in both classical and fuzzy sets)
14 (it results, in general, in another fuzzy set)
16 (they are equal when they have the same membership values for all their elements),
17 (this would be the fuzziest set)

Embedded Fuzzy Applications

1. All fuzzy logic applications are embedded.

2. A unique feature of a fuzzy logic control system is that it uses an error signal to correct the output.

3. Fuzzy control systems are always open-loop systems.

4. A fuzzy logic control system does not require a mathematical model.

5. Feedback could lead to instability.

6. The use of fuzzy logic always leads to superior performance characteristics compared to those of a PID control.

7. Fuzzy logic systems are always stable.

8. IF/THEN rules are unique to fuzzy control systems.

9. Fuzzy logic can solve problems that have no known solutions.

10. Defuzzification is required to reach a crisp action.

11. Embedded systems can work without fuzzy logic.

12. Fuzzy logic could make embedded systems more efficient.

13. In general, the larger the number of rules for an embedded fuzzy control system, the smoother the control surface.

14. The objective of fuzzy control is to replace PID control.

15. A system is stable when its free energy is maximized.

16. Fuzzy logic always leads to optimum solutions.

17. Fuzzy logic could be used for signal processing applications.

18. The step response of a fuzzy system can be measured.

19. Fuzzy control is the only possible embedded application of fuzzy logic.

20. Fuzzy logic = computing with words.

Answers:

False:

1 (not only that, but also there are numerous non-engineering applications),
2 (all feedback control systems generate error signals),
3 (always feedback systems),
6 (not always),
7 (not always, stability needs to be tested and assured),
8 (for example, systems that use look-up tables. What is unique is the allowance and making use of the firing of more than one rule at a time),
14 (fuzzy logic can fine tune PID systems and control systems that have no mathematical model),
15 (when minimized),
16 (no mathematical solution, but an expert must know how to solve the problem so that rules can be generated leading to a linguistic model),
19 (other applications are possible, e.g. signal processing).

Neural Networks

1. Neural networks are distributed computational systems.

2. Neural systems don't degrade gracefully as opposed to von Neumann computing systems.

3. Learning in neural networks occurs by adjusting the synaptic weights.

4. Hebbian learning implies increasing the effectiveness of active connections.

5. All neural networks require training patterns.

6. The structure of neural networks, except ART networks, follows the von Neumann architecture.

7. The XOR problem can be solved by a single neuron.

8. The AND problem can be solved by a single neuron.

9. A multi-layer perceptron is a self-organizing network.

10. Hopfield networks are not self-organizing.

11. Backpropagation is the training algorithm typically used for Kohonen networks.

12. Over-learning leads to learning the noise associated with the input patterns.

13. The hidden layer in a multilayer perceptron represents its memory.

14. The multilayer perceptrons solved the stability-elasticity problem.

15. A hyper-cube is a cube in an n-dimensional space.

16. The stability-elasticity problem is particularly severe in ART networks.

17. The lowest three neurons in a multilayer perceptron are referred to as the basins of attraction.

18. The synaptic weights are usually expressed in milligrams.

19. All neural networks have at least one hidden layer.

20. A Boltzmann network requires hardware systems that can operate at temperatures above room-temperature.

21. Neural networks and fuzzy logic have some complementary characteristics.

Answers:

False:

2 (they do because of the parallel multi-connection architecture)
5 (self-organizing networks do not)
5 (they are parallel processors, digital computers do)
7 (it is not a linearly separable problem)
11 (they are self-organizing)
13 (the synaptic weights represent a distributed memory)
14 (ART did)
16 (on the contrary!)
17 (they are minima in the energy landscape)
18 (that is silly!)
19 (for example, Kohonen networks do not have any),
20 (that is also silly!)

Hybrid Systems

1. Fuzzy systems can learn by example.

2. Neural networks cannot interpret a linguistic statement.

3. There is nothing common between neural networks and fuzzy logic.

4. Neuro-fuzzy and fuzzy-neuro systems represent the zenith in computing technology. No further advances are possible.

5. The aggregation process is a summation operation in the fuzzy neuron model.

6. Fuzzy neural networks that use FNs are typically inhomogeneous networks.

7. ARTMAP is also known as predictive ART.

8. The generalization to learning both analog and binary input patterns is achieved by replacing the classical INTERSECTION operation in ART1 by the fuzzy MAX operation.

9. Fuzzy ARTMAP can operate on input patterns of continuous values whether or not they represent fuzzy sets.

Answers:

False:

1 (neural networks can, fuzzy systems require rules),
3 (both are inspired by the human computational abilities),
4 (certainly not, progress goes on!),
5 (MIN or MAX operations),
8 (replaced by the fuzzy MIN).

Hardware Realization

1. In general, hardware implementation of a fuzzy controller leads to faster operation than implementation using software.

2. In digital implementation of fuzzy systems the term FLIPS refers to a famous Dutch electronics manufacturer.

3. Analog VLSI circuits for fuzzy logic implementation are now obsolete.

4. An ideal current source has an infinite output resistance.

5. An ideal voltage source has an infinite output resistance.

6. Current mirrors are basic building blocks in current-mode circuit implementations.

7. Current mirrors could be used to increase the fan-out of a current-mode circuit.

8. Voltage-mode implementations do not require resistors.

9. Current-mode implementations do not require resistors.

10. It is possible to implement VLSI mixed-mode circuits on the same chip.

11. A bounded difference operation can be implemented with a current mirror and a diode.

12. Only bipolar technology can implement bounded difference operations.

13. Bounded difference circuits can be used to implement fuzzy AND and fuzzy OR functions.

14. An ideal OTA has an infinite input resistance and an infinite output resistance.

15. An ideal Op Amp has an infinite input resistance and zero output resistance.

16. There are numerous reliable analog memory modules.

17. A fuzzy flip-flop can be thought of as a basis for fuzzy memory modules.

18. OTAs cannot be used to implement the bounded difference operation.

19. An Op Amp summing amplifier circuit can implement the basic neuron model.

Answers:

False:

2 (refers to Fuzzy Logic Inferences per Second),

3 (on the contrary, it is gaining more popularity),

5 (an ideal voltage source has zero output resistance so that the load does not affect the value of the output voltage),

8 (it does and hence more power consumption),

12 (CMOS technology also can be used),

16 (No, the lack of reliable analog memory modules is the main weak point of analog implementation),

18 (they can, it is a basic building block of OTA-based implementation of fuzzy logic operations).

Genetic Algorithms

1.   Introduction

Genetic algorithms, GAs, are general-purpose search and optimization procedures. They were inspired by the biological evolution principle of *survival of the fittest*. This led to the metaphoric use of terminology borrowed from the field of biological evolution. For example, *genes* and *chromosomes* in the context of GAs refer to binary strings of 1s and 0s (integer and real numbers are less commonly used) that represent the data to be processed. GAs do not deal directly with the parameters of the problem to be solved. They work with codes which represent the problem and produce codes which represent the solution.

The fundamental idea is to maintain a population of individuals (knowledge structures) that evolves over time through a process similar to birth and natural selection, i.e., the generation of new individuals (children) by combining the structures of two individuals (parents). Each structure in the population represents a potential solution to the problem at hand. Competition based on fitness leads to the formation of new, fitter individuals. Genetic operators, such as crossover, and mutation, are used to generate the new ones.

GAs were reported to have achieved great success because of their ability to exploit accumulated information about an initially unknown search space leading to moving subsequent searches into a more useful subspace. GAs are also flexible tools and can be used in combination with other techniques including fuzzy logic and neural networks. For example, fuzzy logic techniques lead to a solution to a control system problem, but not necessarily the optimum solution. A fuzzy knowledge base can be encoded as an individual in a genetic algorithm population. A GA can then be used to produce the best knowledge base for the desired objectives. GAs have also been applied to membership function adjustment. In section 2.8 of the appendix, numerous sources are cited that detail applications that combine fuzzy logic and genetic algorithms.

The main structure of a genetic algorithm is illustrated in the simplified flow chart of Figure A.1.

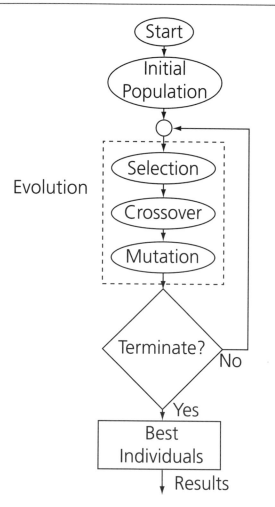

Figure A.1: Simplified flow chart of a genetic algorithm.

The steps of the flow chart could be briefly explained as follows:

Initial Population

The initial population is created randomly for problems about which no prior knowledge exists. The population size has to be large enough to ensure sufficient diversity to cover the possible solution space. Numerous problems start with 100 individual genes.

Evolution

The evolution of a population in genetic algorithms follows the Schema Theorem put forward by Holland (J. H. Holland, *Adaptation in Natural and Artificial Systems*, University of Michigan Press, 1975). A schema represents a subset of the population composed of individuals with similar bits in certain positions. A schema is characterized by its defining length (number of bits between the first and last bits with fixed values). The schema: 1*1*, for example, describes the set of individuals whose first and third bits are 1s. The symbol * indicates that any value, whether 1 or 0, is acceptable. Evolution proceeds as a result of applying a number of, but not necessarily all of the genetic operators: SELECTION, CROSSOVER, MUTATION, and INVERSION.

SELECTION

Selection is the process of extracting individuals from the existing population whose fitness values are relatively higher than the rest. There are numerous methods used for selection. Two commonly used methods are:

The so-called Roulette Selection (also known as stochastic or proportional selection) where the weighting of the fitness value of each individual is calculated as a percentage of the average fitness of the entire population. The second is referred to as Tournament Selection where individuals are selected randomly and then the ones with the highest fitness are chosen.

CROSSOVER

Crossover or recombination is a variation operator. It merges the genetic information of two existing individuals (parents), picked up by the selection operation, to create two new individuals (children). There are numerous ways in which such an operation could be achieved. The simplest is the one-point crossover where two selected individuals are cut at a randomly selected point. The tails (the parts after the cutting point) are swapped, leading to two new individuals. This is illustrated in Figure A.2.

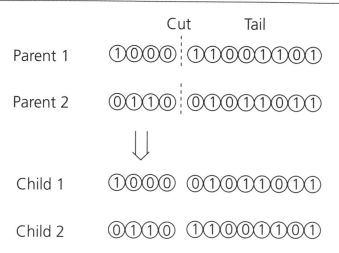

Figure A.2: One-point crossover operation.

MUTATION

Mutation here means choosing at random bits in the knowledge structure and flipping them according to a specified rate. The mutation operator forces the algorithm to search new areas. It helps avoiding premature convergence and finding the global optimal solution.

INVERSION

This operator is used in problems such as the traveling salesman problem and layout problems. In that operation, two points are randomly selected from an individual and the part of the string between them is inverted.

Termination

The objective is to test whether the optimization criteria are met or not. The point at which the process terminates needs to be defined. Termination criteria set by the user include:

- Target fitness value

- Maximum number of generations

- Equal number of individuals

- Maximum execution time

The next section provides a classified Web bibliography of genetic algorithms. All cited work was made available by the authors through the Web at no charge. It is hoped that the selection and the classification will save the readers time in exploring the area of genetic algorithms. The bibliography starts with glossaries that are authored by various sources. A few excellent tutorials with varied approaches are cited next. They are followed by a small listing of research papers in the area. A selection of interactive Java applets that can run through the Web is also presented. The reader can experiment with them to gain further insight into genetic algorithms operations. They are followed by a selection of sites that provide GA codes in numerous languages including Java, MATLAB, C, C++, and Perl. The final section of the bibliography provides links to references on the combination of fuzzy logic and genetic algorithms.

2. Classified Annotated Web Bibliography

This section provides resources accessible freely through the Web. They are classified into eight groups: Glossaries, Introductions and Tutorials, Research Papers, Interactive Demos, Java, MATLAB, and other languages resources. The last group provides resources for the applications of genetic algorithms combined with fuzzy logic.

2.1 Glossaries

Glossary of Evolutionary Algorithms

http://ls11-www.cs.uni-dortmund.de/people/beyer/EA-glossary/
def-engl-html.html

An extensive stand-alone glossary in HTML format provided by Hans-Georg Beyer (University of Dortmund, Dept. of Computer Science, Dortmund, Germany), Eva Brucherseifer, Wilfried Jakob, Hartmut Pohlheim, Bernhard Sendhoff, and Thanh Binh To. The glossary gives detailed definitions for numerous terms, and the definitions have hypertext for ease of further referencing.

GA Glossary

www.santafe.edu/sfi/publications/Bookinforev/aiinep-gloss.html

A glossary from *Adaptive Individuals in Evolving Populations: Models and Algorithms* edited by R. K. Belew and M. Mitchell, Santa Fe Institute, Addison-Wesley, Reading, MA, 1996. It also provides a set of references on the topic.

A Short GA Glossary

www.cs.ucl.ac.uk/staff/W.Langdon/FOGP/glossary.html

A Glossary from *Foundations of Genetic Programming* by W. B. Langdon and R. Poli, School of Computer Science, The University of Birmingham, Edgbaston, Birmingham UK, Kluwer Academic Publishers, Boston,1998. It is a concise, short glossary that defines about 45 terms.

2.2 Introductions and Tutorials

Introduction to Genetic Algorithms

www.rabatin.com/papers/gaintro/ga_intro.html

A simple introduction to genetic algorithms by Arthur Rabatin, Structured Derivatives Group of EdF Trading Ltd. Topics covered include:

Overview, Representation of the Problem Domain, Fitness Value, Learning Process Initialization and Evolving Structure, and Selecting GA Process Parameters.

GA Short Introduction

http://sun1000.pwr.wroc.pl/./AMIGA/ARTech/i002/GeneticAlgorithms_1.html

A short introduction by Marcel Offermans, Mechanical Engineering Department, Delft University of Technology, The Netherlands. The main topics discussed include:

- What is a genetic algorithm?
- What can you use it for?
- How to create the chromosomes?
- What does the genetic algorithm do?

How the Genetic Algorithm Works

www.pmsi.fr/gainita.htm

An HTML document that gives a brief description of how the Genetic Algorithm works and some of its applications. It is provided by PMSI Saxon et Galvano. Topics covered include:

What the Genetic Algorithm is useful for, How the Genetic Algorithm works, Why the Genetic Algorithm is a good idea, The Genetic Algorithm and data mining, and Why this technique is interesting.

Several software that demonstrate the operation of GAs are available for downloading, they include:

- FOOD, Robots learning to look for food
- TSPGA, Travelling Salesman's Problem solution
- GAFUNC, various function optimization problems
- TSPSA, simulated annealing (This freeware is dedicated to all humble, friendly scientists and also to Professor Robert Azencott).

Genetic Algorithms

www.ifs.tuwien.ac.at/~aschatt/info/ga/genetic.html

An introduction to Genetic Algorithms by Alexander Schatten, Institut für Oftwaretechnik, Vienna University of Technology. The document is in HTML format and it gives a good introduction to the topic with an application example.

Genetic Algorithms

http://samizdat.mines.edu/ga_tutorial/

A tutorial by Darrell Whitley, Computer Science Department, Colorado State University.

It covers the canonical genetic algorithm as well as more experimental forms of genetic algorithms, including parallel island models and parallel cellular genetic algorithms. It is available for downloading in postscript format (400Kb).

Genetic and Evolutionary Algorithms: Principles, Methods and Algorithms

www.geatbx.com/docu/algindex.html

A tutorial provided by GEA Toolbox. Topics include:

Selection: Rank-based fitness assignment, Roulette wheel selection, Stochastic universal sampling, Local selection, Truncation selection, Tournament selection, Comparison of selection schemes, Recombination: Real valued recombination, Binary valued recombination (crossover), Mutation: Real valued mutation, Binary mutation, Reinsertion: Global reinsertion, Local reinsertion, Parallel implementations, and Migration: Global model, and Diffusion model.

Genetic Algorithms Applications

www.doc.ic.ac.uk/~nd/surprise_96/journal/vol4/tcw2/report.html

An introduction to genetic algorithms by Naranker Dulay, Department of Computing, Imperial College, London. Topics include: Brief Overview, Who can benefit from GA, Applications of Genetic Algorithms, and Artificial Life.

Evolutionary Algorithms Notes

www.arch.usyd.edu.au/~mike/GA/SLIDES/GA.htm

Introduction to evolutionary algorithms provided by Michael Rosenman, Department of Architectural and Design Science, The University of Sydney. Topics include: Evolutionary Algorithms, General Methodology, Nature of EAs, Terminology and Comparison, Genotype Representation, Genetic Operators, Selection, Fitness Function, Constraints, and Examples.

The Genetic Programming Notebook
www.geneticprogramming.com/

This is a comprehensive site that provides a link to a concise clear tutorial on GA, in addition to links to: Bibliographies, Books, Conferences, GA Courses, GA FAQ, Research Groups, GA Journals, GA Papers, Repositories, and GA related software.

Optimization of Structuring Elements by Genetic Algorithms
http://ltswww.epfl.ch/pub_files/brigger/thesis_html/node58.html

A section from a PhD thesis by Patrick Brigger, the Swiss Federal Institute of

Technology/Lausanne (EPFL). The thesis title is *Morphological Shape Representation Using the Skeleton Decomposition: Application to Image Coding.* The section discusses:

- Basic principles of the genetic algorithm
- Transposition to the optimization of the structuring element
- Reproduction plan
    - Roulette-wheel reproduction
    - Steady-state reproduction
    - Simplex reproduction
- Comparison of reproduction algorithms
- Genetic operators
- Evaluation of the importance of the genetic operators
- Shape oriented crossover operator.

An Introduction to Genetic Algorithms for Numerical Optimization
www.hao.ucar.edu/public/research/si/pikaia/tutorial.html

This site gives access to an advanced tutorial by Paul Charbonneau and is available in postscript format. The tutorial includes exercises that require the use of the genetic algorithm-based optimization subroutine PIKAIA, which is also available for downloading.

Genetic Algorithms: Concepts and Designs
http://sant.bradley.edu/ienews/98_4/GENET2.pdf

An article that introduces GAs and their role in control systems. The article has 12 references and appeared in the IEEE Industrial Electronics News Letter, 45, 4, 1998.

2.3 Research Papers

Genetic Algorithms
www.arch.columbia.edu/DDL/cad/A4513/S2001/r7/

A short treatise by John H. Holland that outlines the fundamental ideas of genetic algorithms.

A New Schema Theory for Genetic Programming
http://cswww.essex.ac.uk/staff/poli/papers/Poli-GP1997.pdf

A paper by R. Poli and W. B. Langdon, The University of Birmingham. It reviews the results obtained in the theory of schema in genetic programming. It also proposes a new, simpler approach for genetic programming.

A List of Papers
http://citeseer.nj.nec.com/MachineLearning/GeneticAlgorithms/hubs.html

This site provides links to a large number of research papers related to GAs.

GA Archives
www.aic.nrl.navy.mil/galist/

The Genetic Algorithms Archive is a repository for information related to research in genetic algorithms and other forms of evolutionary computation. It is maintained by Alan Schultz, The Navy Center for Applied Research in Artificial Intelligence, USA. Its contents include: Calendar of EC-related events, GA-List Archive: back issues, source code, and information, Links to other GA-related information, Links to EC research groups' home pages, and Dissertations on Evolutionary Computation.

2.4 Interactive Demos

GA Interactive Experiment
www.oursland.net/projects/PopulationExperiment/

An applet designed by Allan Oursland, the University of Texas at Austin. This applet demonstrates a continuous value genetic algorithm on a variety of problem spaces with a variety of reproduction methods. Its objective is to help visualize the processes in a genetic algorithm. A detailed explanation along with the source code is provided.

Test Environment
www.gawalk.com/

An applet for experimenting with GAs. Extensive documentation and the software are available to the user.

Interactive Genetic Algorithm Demonstrator
http://userweb.elec.gla.ac.uk/y/yunli/ga_demo/

A demo applet provided by Dr. Li, Department of Electronics and Electrical Engineering, University of Glasgow, Glasgow, UK.

The GA Playground
www.aridolan.com/ga/gaa/gaa.html

The GA Playground is a general purpose genetic algorithm toolkit provided by Ariel Dolan. It is implemented in Java. Extensive documentation along with examples and test problems are provided.

Genetic Java
http://www4.ncsu.edu/eos/users/d/dhloughl/public/stable.htm

The site is provided by Dan Loughlin, Department of Civil Engineering, North Carolina State University, USA.

GA Optimizer
http://ai.bpa.arizona.edu/~mramsey/ga.html

An interactive applet with documentation provided by Marshall Ramsey, University of Arizona, UAMIS AI Group. The source code is available for download.

2.5 Java Related Resources

GA in Java
www.esatclear.ie/~rwallace/gp.html

The site enables the downloading, in zip format, of two public domain applications created by Russell Wallace to implement genetic programming.

ECJ, Evolutionary Computation in Java
www.cs.umd.edu/projects/plus/ec/ecj/

The site provides a research-oriented Evolutionary Computing Java program, ECJ. It is authored by Sean Luke, Department of Computer Science, George Mason University, USA. Documentation and source code are available for download.

MathTools
www.mathtools.net/MATLAB/Genetic_algorithms/Java/

An annotated list of Java applets related to GA provided by MathWorks.

2.6 Matlab Related Resources

GEATbx:
www.systemtechnik.tu-ilmenau.de/~pohlheim/GA_Toolbox/
www.geatbx.com/

Genetic and Evolutionary Algorithm Toolbox for use with MATLAB. The M-files are not free, and the documentation is available for download along with tutorial examples.

GA M-files
ftp://ftp.mathworks.com/pub/tech-support/solutions/s1248/genetic/

A collection of GA M-files, written by Andrew Potvin, the MathWorks, Inc. It is available for free download.

MATLAB for solving GA problems
www.pmfst.hr/ceepus/radovi/Orlikova.pdf

A paper by Sona Orlikova, Brno University of Technology, Czech Republic. It introduces GA and its implementation in MATLAB. It appeared in the Proceedings of the Fifth Symposium on Intelligent Systems, Split, Croatia, 2002.

Relation of Fuzzy-Adaptive Genetic Algorithms in a MATLAB Environment
http://phobos.vscht.cz/matlab01/matousek.pdf

A paper by R. Matoušk, Institute of Automation and Computer Science, Brno University of Technology, Czech Republic. It discusses integrating fuzzy logic and genetic algorithms. MATLAB Fuzzy Logic Toolbox was used for the fuzzy inference system part.

2.7 Other Languages Resources

GAlib
http://lancet.mit.edu/ga/

A free C++ Library of Genetic Algorithm Components developed by Matthew Wall,

Massachusetts Institute of Technology. It includes tools for using genetic algorithms to perform optimization in any C++ program using any representation and genetic operators. Documentation is available, and it includes an explanation of how to implement a genetic algorithm with examples.

GA Source Code Collection
www.aic.nrl.navy.mil/galist/src/

This site provides links to source code for implementations of genetic algorithms and other EC methods. The software is listed alphabetically, and by language. It is maintained by Alan Schultz, Naval Research Laboratory, USA.

Lithos Evolutionary Computing
www.esatclear.ie/~rwallace/lithos.html

Lithos is an evolutionary computation system. It is provided in the public domain by Russell Wallace. The C++ source code and Windows executable files are provided along with detailed documentation.

Diophantine Equation Solver

www.generation5.org/diophantine_ga.shtml

A C++ program that solves a diophantine equation using genetic algorithms provided by Generation5.org. Detailed explanations and examples of the code are also provided.

Advanced Examples of Genetic Algorithms with Perl
www-106.ibm.com/developerworks/linux/library/l-genperl2/
?t=gr,lnxw961=nextgen

An article by Teodor Zlatanov, Gold Software Systems, that discusses advanced aspects of GAs in Pearl. It builds on what was discussed in a previous article titled "Genetic Algorithms Applied with Perl", www-106.ibm.com/developerworks/linux/library/l-genperl/

2.8 Fuzzy Logic and Genetic Algorithms

Genetic Algorithms and Fuzzy Logic in Control Processes
http://citeseer.nj.nec.com/cordon95genetic.html

A comprehensive discussion of the application of genetic algorithms in control systems by O. Cordón, F. Herrera, and M. Lozano, Department of Computer Science and A.I., University of Granada, Spain. The report reviews fuzzy logic controller concepts and applications, discusses the fundamentals of genetic algorithms, and details the design of fuzzy logic controller using genetic algorithms. It also discusses learning classifier systems. References up to the year 1995 are included.

Combination Fuzzy Logic-Genetic Algorithms Bibliography
http://decsai.ugr.es/~herrera/fl-ga.html

An HTML document that presents a classified listing of papers on the combination of fuzzy logic and genetic algorithms with a bibliography by O. Cordón, F. Herrera, and M. Lozano, Department of Computer Science and A.I., University of Granada, Spain. It has over 500 references up to the year 1997, but the texts of the papers are not accessible. The categories include: Fuzzy Genetic Algorithms, Fuzzy Clustering, Fuzzy Optimization, Fuzzy Neural Networks, Fuzzy Relational Equations, Fuzzy Expert Systems, Fuzzy Classifier Systems, Fuzzy

Information Retrieval, Fuzzy Decision Making, Fuzzy Regression Analysis Fuzzy Pattern Recognition and Image Processing, Fuzzy Classification – Concept Learning Fuzzy Logic Controllers, Fuzzy Logic-Genetic Algorithms Framework, and Fuzzy Logic Miscellaneous. See also by the same authors in pdf format: *A Classified Review On The Combination Fuzzy Logic-Genetic Algorithms Bibliography*: 1989-1995 at: http://citeseer.nj.nec.com/cordon95classified.html

Fuzzy Logic and Genetic Algorithms for Intelligent Control
www.csu.edu.au/ci/vol02/mm94n2/mm94n2.html

A paper in HTML format by M. Mohammadian (Department of Computer Science Edith Cowan University, Perth, Western Australia) and R. J. Stonier (Department of Mathematics and Computing, Central Queensland University, Rockhampton, Australia).

It presents the control of a mobile robot in the presence of an obstacle using a system that integrates a fuzzy logic control system and genetic algorithms. The paper appeared in *Complexity International*, 2, April, 1995.

Evolutionary Learning in Fuzzy Logic Control Systems
www.csu.edu.au/ci/vol03/stonier/stonier.html

A paper by R. Stonier and M. Mohammadian which appeared in Complexity International, April 1996. It is available in HTML format. The authors discussed the use of genetic algorithms to learn fuzzy rules in fuzzy logic control systems. The system is applied to collision-avoidance problems and target-tracking for mobile robots. Application to the control of traffic flow approaching a set of intersections and interest rate prediction was also briefly discussed.

Dynamic Control of Genetic Algorithms using Fuzzy Logic Techniques
http://citeseer.nj.nec.com/lee93dynamic.html

A paper by M. A. Lee and H. Takagi, University of California (Davis and Berkeley), USA. It appeared in the *Proceedings of the Fifth International Conference on Genetic Algorithms*, ICGA'93, Urbana-Champaign, Il, 1993, pp. 1293-1296. The authors proposed and experimented with using fuzzy logic techniques to dynamically control parameter settings of genetic algorithms. They described a Dynamic Parametric GA: a GA that uses a fuzzy knowledge-based system to control GA parameters. They introduced a technique for automatically designing and tuning the fuzzy knowledge-base system using GAs.

Design of Sophisticated Fuzzy Logic Controllers using Genetic Algorithms
www.mech.gla.ac.uk/Research/Control/Publications/Rabstracts/abs94001.html

A paper by K.C. Ng and Y. Li, Faculty of Engineering, Glasgow University, Scotland. It appeared in the *Proceedings of the Third IEEE International Conference on Fuzzy Systems*, Orlando, FL, 1994, pp. 1708-1712. The authors developed genetic algorithms for automatic design of high performance fuzzy logic controllers. The controller design space is coded in base-7 strings (chromosomes), where each bit (gene) matches one of the 7 discrete fuzzy values. The developed approach is subsequently applied to the design of a PI- type fuzzy controller for a nonlinear water level control. The paper is available in PostScript format.

GA and Fuzzy Logic for Plant Control
www.supelec-rennes.fr/rennes/si/equipe/lme/PUBLI/ifsicc.pdf

A paper by N. Perrot et al. which appeared in the Proceedings of the *International Conference on Fuzzy Systems and Intelligent Controls Conference*, IFSICC'96, Maui, Hawaii, April, 1996. The authors described a fuzzy controller used for the automatic control of the crossflow microfiltration process of raw cane sugar. The fuzzy controller was optimized off line using genetic algorithms and neural networks.

Mixing Fuzzy, Neural and Genetic Algorithms
http://polimage.polito.it/~marcello/articoli/systems.man.pdf

A paper by M. Chiaberge, et al. It appeared in the Proceedings of the *IEEE Conference on Systems, Man and Cybernetics*, Vancouver, B.C., October 1995, pp. 2988--2993. The authors outline the merits of hybrid neural and fuzzy controllers and introduce the concept of Hierarchical Hybrid Fuzzy Controllers. Genetic algorithms, role in control systems is explained and used for parameter optimization.

Neural Chips and Evolvable Hardware
http://glendhu.com/ai/neuralchips/

This page provides links related to neural chips, including:

- Vision Chips, Caltech: Chips modeled on the visual system of the fly. The chip contains a photosensitive array and processing circuitry all in one.

- Morphing brains to silicon, University of Pennsylvania: neuromorphic engineering techniques to build microprocessors closely modeled on biological nervous systems.

- ETL, Japan: A research group developing a self-reconfigurable neural network chip.

- The Learning Processor, Axeon Ltd: A company developing a microprocessor which contains 256 RISC-like processors in parallel on one chip.

Symbols and Acronyms

$\{x_1, x_2, x_3 \ldots\}$	A set of elements $x_1, x_2, x_3 \ldots$
$\{x \vert p(x)\}$	A set determined by property p
$[x_{ij}]$	Matrix
$\mathbf{X}$	Matrix
$[a,b]$	Closed interval of real numbers between a and b
$[a,b)$	Interval of real numbers, closed in a and open in b
$[a,\infty)$	Set of real numbers greater than or equal to a
$x \in X$	Element x belongs to set X
$x \notin Y$	Element x does not belong to set Y
$A \subset B$	A is a subset of B
$A \not\subset B$	A is not a subset of B
$\varnothing$	Empty set
$\vert A \vert$	Cardinality of set A
$A \cup B$	Union of sets A and B
$\max[x_1, x_2, \ldots]$	Maximum value of $x_1, x_2, \ldots$
$\vee$	Maximum of
$A \cap B$	Intersection of sets A and B
$\min[x_1, x_2, \ldots]$	Minimum value of $x_1, x_2, \ldots$
$\wedge$	Minimum of
$\neg A$	Complement of set A, also $\overline{A}$
μ_A	Membership function of fuzzy set A
$\mu_{\overline{A}}$	Membership function of the complement of fuzzy set A

A_α	a-cut of set A
supp A	Support of fuzzy set A
$\ominus$	Bounded difference
$\oplus$	Bounded sum
$A \rightarrow B$	A implies B
$R \circ Q$	max-min composition of binary fuzzy relations P and Q
$x \Leftrightarrow y$	x if and only if y
iff	if and only if
$\forall$	For all
$\exists$	There exists at least one
$\triangleq$	Defined by
$\displaystyle\sum_{i=1}^{n} x_i$	Summation of $x_1 + x_2 + \ldots + x_n$
$\displaystyle\prod_{i=1}^{n} x_i$	Multiplication of $x_1 x_2 \ldots x_n$
β	Beta, common-emitter forward-current amplification factor
g_m	Transconductance
Op Amp	Operational Amplifier
OTA	Operational Transconductance Amplifier
BJT	Bipolar Junction Transistor
MOS	Metal Oxide Semiconductor
FET	Field-Effect Transistor
MOSFET	Metal-Oxide-Semiconductor Field-Effect Transistor
CMOS	Complementary symmetry MOS
BiMOS	Bipolar-MOS
NMOS	n-Channel MOS
IIL	Current Injection Logic
TTL	Transistor-Transistor Logic
PLC	Programmable Logic Controller
PLD	Programmable Logic Device

CPLD	Complex PLD
DSP	Digital Signal Processing (Processor)
VLSI	Very Large Scale Integration
IC	Integrated Circuit
ASIC	Application Specific Integrated Circuit
FPD	Field Programmable Logic
GA	Gate Array, Genetic Algorithm
CBIC	Standard-Cells Based Integrated Circuits
FLC	Fuzzy Logic Control (Controller)
FLIPS	Fuzzy Logic Inferences Per Second
FF	Flip-flop
ANN	Artificial Neural Network (NN: Neural Network)
CAD	Computer-Aided Design
D/A, DAC	Digital-to-Analog Converter
A/D, ADC	Analog-to-Digital Converter
ROM	Read-Only Memory
RAM	Random-Access Memory
SRAM	Static Random Access Memory

Circuit Symbols and MOSFETs

MOSFET is an acronym for Metal-Oxide-Semiconductor Field-Effect Transistor. These devices are constructed with a semiconductor layer that is a wafer of a single crystal of silicon, a layer of silicon dioxide, which is an insulator; and a metallic layer. An example of a physical structure of a MOS transistor is shown in the figure below. The figure shows an n-channel MOS, referred to as NMOS. It is common to use polycrystalline silicon instead of the metal layer, yet the term MOS is still in use. A more general term that is sometimes used is IGFET (insulated gate FET). The operation of the device is based on controlling the conductance of the channel through the voltage applied at the gate. There are two types of MOSFET: enhancement and deletion. In the enhancement-type, it is arranged that the gate voltage can enhance the conductance by inducing more mobile charge carriers. In the depletion-type it is arranged that the gate voltage can reduce the conductance by depleting the channel of mobile charge carriers. These possible variations of the MOS devices have led to a variety of circuit symbols being used. Furthermore, varied symbols are sometimes used to mean the same thing contrary to the case of the BJT, where there are only two symbols. This situation could confuse the uninitiated in trying to follow the various VLSI schemes of implementing fuzzy logic and neural networks. Here, various symbols that appeared in the cited literature are listed. Still, the reader may find some minor variations in some of the literature. The following list is hoped to ease this difficulty.

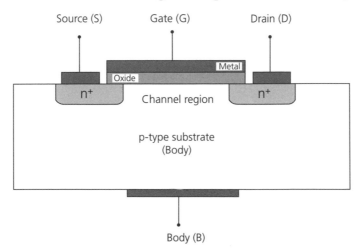

Physical structure of enhancement-type NMOS transistor.

n-Channel Enhancement Type

Standard

Simplified with substrate connected to source

<u>Common for discrete circuits</u>

<u>Common for integrated circuits</u>

Other symbols in use

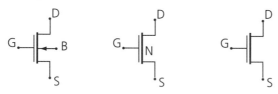

p-Channel Enhancement Type

Standard

Simplified with substrate connected to source

Common for discrete circuits   Common for integrated circuits

Other symbols in use

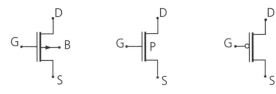

n-Channel Depletion Type

Standard

Simplified with substrate connected to source

Common for discrete circuits   Common for integrated circuits

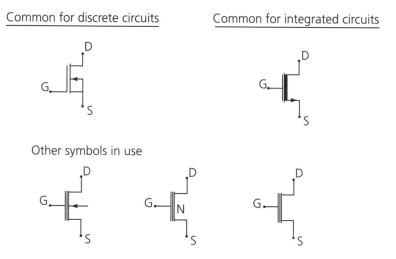

Other symbols in use

p-Channel Depletion Type

Standard

Simplified with substrate connected to source

Common for discrete circuits   Common for integrated circuits

Other symbols in use

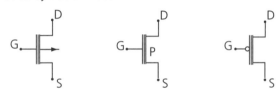

Glossary

ADALINE: An acronym for Adaptive Linear Element. An early neural network put forward by Bernard Widrow and Ted Hoff in the 1960s. MADALINE is an array of MANY ADALINE.

ADAPTIVE: A system that can be modified during operation to meet specified criteria. Adaptation is sometimes used synonymously with learning.

ADAPTIVE FUZZY SYSTEM: A fuzzy system that does not require rules from a human expert; it generates and tunes its own rules.

ALGORITHM (ALGORISM): A step-by-step procedure or a precisely defined set of rules that provide a solution to a problem in a finite number of steps. Alkhawarizmi (780-850 AD) wrote: *with my two algorithms, one can solve all problems – without error, if God will.*

ANCIENT IF/THEN SYSTEM: A papyrus (~3000 BC) purchased in a Luxor (Egypt) antique shop by the Egyptologist Edwin Smith in 1882, that has 48 surgical algorithms for the treatment of neural injuries. The algorithms have the form of: symptom-diagnosis-treatment-prognosis expressed in IF/THEN format.

ANTECEDENT: The clause that implies the other clause in a conditional statement. The initial, or the IF part of a fuzzy rule.

ARGUMENT: In logic, an argument is not a disagreement, but a piece of reasoning with one or more premises and a conclusion. Arguments are usually divided into two kinds, deductive and inductive.

An argument is sound iff it is valid and all of its premises are true. It is valid iff it is not possible to have all true premises and a false conclusion. Common valid arguments include:

MODUS PONENS
> If p then q, p; therefore q.

MODUS TOLLENS
> If p then q, not-p; therefore not-q.

ART: Adaptive Resonance Theory: a self-organizing network. It was introduced by Stephen Grossberg and Gail Carpenter. The first version, ART1, can process only binary input patterns. The second version, ART2, can process real input patterns. ART3 is an improved ART2 in which the processing is more stable. ARTMAP is a supervised version of ART.

ARTIFICIAL NEURAL NETWORKS (ANN): A parallel distributed computing system inspired by the human nervous system. They are referred to also as Electronic Neural Networks, or simply neural networks if there is no fear of confusion with the biological one; the acronym NN is also commonly used.

ARTIFICIAL STUPIDITY: From a sociological point of view the term was defined by T. R. Yong in his treatise: *Structurally Stupid Societies: Exploration in Artificial Stupidity, Red Feather Institute, No. 155, 1991.* According to Young, a society (and by extension a system) is naturally stupid when it lacks the means to reflect on its own behavior and modify it in ways that ensure the well-being of its members and the natural and social environment upon which it depends for survival. A society is artificially stupid when it has the means to reflect on its own behavior and ensure the well being of its citizens, but fails to do so. In artificially stupid societies, the data are collected, processed, and may even be used. However, such data are mediated by structures that lead to stupidity. To be stupid, one must think locally and act globally. Yong further defined a stupidity quotient, SQ test. He also considered the advertising and public relations industries, where resources are used to manufacture images in lieu of quality, to be the largest single body of experts designing and promoting artificial stupidity. He also suggested that ignorant systems do not have access to information; or having access to information, do not have the means to process it. Dumb societies are those in which most people are rendered silent. Young's approach does not lead to applications. One could say that, from an engineering point of view, that artificial stupidity is the modeling of naturally or artificially stupid systems (in Young's sense) with the objective of predicting realistic outcomes. For example, a system designed to emulate the behavior of a human expert in a particular area could be used to predict the wrong decisions the expert would make outside the field of expertise.

ASSEMBLER: A program that creates object code from a symbolic description of instructions.

ASSOCIATIVE MEMORY: A system that stores data in parallel and recalls them based on some feature of the data.

ASYNCHRONOUS: An event not coordinated with a clock.

Axiom: For mathematicians, an axiom is a statement or proposition that is stipulated to be true and it is a convenient starting point. An assumption is a statement that is taken to be true for the purpose of a particular argument, but may not otherwise be accepted.

Axon: The output connection of the biological neuron over which signals are sent to other neurons.

Backpropagation: A supervised learning algorithm for multilayer perceptrons. It operates by calculating the value of the error function for a known input, then backpropagating the error from one layer to the previous one. Each neuron has its weights adjusted so that it reduces the value of the error function until a stable state is reached.

Basins of Attraction: The valleys of the energy surface of a neural network. The energy surface is a graph of the energy function vs. weights, with the energy function being a measure of the amount by which the input differs from the desired output. Thus, the basins of attraction give all possible solutions, i.e., values of weights that produce correct output for a given input.

BIOS: Basic Input/Output System. Low-level operating software in a computer system.

It is the first program that runs on a PC each time it is turned on. It first determines which peripherals are in place and operational. Then it loads the operating system or its crucial parts into RAM from the hard disk or disk drive.

Boltzmann Machine: A neural network algorithm that is based on statistical mechanics. It uses simulated annealing to reach stable states.

Cache Memory: A small memory that holds copies of frequently-accessed memory locations for fast access.

Cardinality: The cardinality of a set is the number of elements in the set, intuitively the size or magnitude of the set.

CMOS: Complementary metal oxide semiconductor, commonly used in VLSI (Very Large Scale Integration) technology. It is a combination of a n-channel and a p-channel MOS in a single circuit. It has very low power consumption.

Comparator: An electronic circuit that has two inputs and one output. The output is zero, positive, or negative depending on whether the input is equal to, greater than, or smaller than a reference input.

Compensatory Operators: Non-Zadeh operators, i.e., operators not defined by simple min-max rules. They fall into two general categories: the first encompasses operators based on simple arithmetic transformations, such as the bounded sum and difference. The second encompasses operators based on more complex functional transformations, such as Yager operators.

Competitive Learning: A learning algorithm that requires neurons to compete with each other in adjusting their weights. The neuron with the maximum output and its neighbors are allowed to adjust their weights. The concept is sometimes referred to as *winner take all*.

Comprehensive Information Theory: The traditional Information Theory (Shannon Theory) is concerned with formal features of the stimulus, the meaning and values are ignored. Information is defined in terms of negative entropy that is purely statistical in nature. The Comprehensive Information Theory, CIT, put forward by Y. X. Zhong in the 1990's addresses this deficiency in Shannon Theory. CIT is defined as being composed of three factors: syntactic, semantic, and pragmatic. The syntactic information on the stimulus is concerned with its formal factor, the semantic with the meaning factor, and the pragmatic with the value factor. The syntactic information could be either statistical or fuzzy depending on the stimulus considered. The Comprehensive Information Theory can in general be expressed in a matrix of the form: [X, C, T, U]. Shannon theory can be considered a special case of CIT.

Conclusion: The result of an argument or inference.

Connectionism: Viewed by some as a movement in cognitive science to model the brain based on interconnection of many simple units forming an artificial neural network to produce complex behavior. The history of connectionism, however, spans a wide range of disciplines over a number of centuries. Modern philosophers are interested in connectionism because of its promise to provide an alternative to the view that the mind is similar to a digital computer processing symbolic language. However, dynamic systems theorists and symbolicists have put forward numerous critiques of connectionism. Landmarks in the modern history of connectionism include: the seminal contribution by Waren McCulloch and Walter Pitts, *A Logical Calculus of Ideas Immanent in Nervous Activity* published in 1943; the discovery of the so-called *perceptron convergence procedure* to train a two-layer network by Frank Rosenblatt in 1960's; the critique of neural networks by Marvin Minsky and Seymour Papert published in 1969 which is considered a major cause for the decline in neural networks research during the 1970's; the work of Geoffrey Hinton, James McCelland, David Rumelhart, Paul Smolensky, and other members of the parallel Distributed Processing Research Group in the 1980's to which the revival of the popularity of neural networks is attributed.

CONNECTION WEIGHT: In an artificial neural network the connection weight simulates the strength of the synaptic connection to a neuron. Learning leads to correctly adjusting the connection weights in the network.

CONSEQUENT (SUCCEDENT): The resultant clause in a conditional statement. The final or the THEN part of a fuzzy rule.

CO-PROCESSOR: An optional unit added to a CPU that is responsible for executing some of the CPU's instructions.

CRISP SET: A set that does not allow degrees of membership; an item is either a member or not. It is also referred to as a classical set. All sets are subsets of the universal set (mother of all sets).

DATA CLUSTERING: A technique in which objects that have similar characteristics form a cluster. The criterion for determining similarity is implementation dependent. Clustering is different from classification where objects are assigned to pre-defined classes. In clustering, the classes are to be defined. Data clustering could increase the efficiency of a database if the information that is logically similar is physically stored together.

DATA MINING: The process of discovering meaningful and useful new correlations, patterns, or trends in large amounts of data using pattern recognition techniques including neural networks.

DEFUZZIFICATION: The process of determining the best crisp representation of a given fuzzy set.

DEGREE OF MEMBERSHIP: An expression of the confidence or certainty that an element belongs to a fuzzy set. It is a number that ranges from zero to one. Membership degrees are not probabilities and they do not have to add up to one.

DIGITAL SIGNAL PROCESSOR (DSP): A microprocessor whose architecture is optimized for digital signal processing applications.

DILEMMA: In logic, a dilemma is a choice in which there are only two options. One can refute a dilemma by finding a third possibility.

EMBEDDED COMPUTER SYSTEM: A computing system used to implement some functionality other than that of general-purpose computers, such as a laptop or a desktop computer. A distributed embedded system is an embedded system built around a network.

EPROM: Erasable PROM; a programmable read-only memory that can be erased and reprogrammed.

EXCLUDED MIDDLE LAW: The principle that every proposition is either true or false. The principle leads to classical set theory. Fuzzy logic and fuzzy sets do not obey this law, since fuzzy sets allow partial membership.

FIRING STRENGTH: The degree to which the antecedent of a particular fuzzy rule is satisfied. It determines the degree to which the rule constitutes the net outcome.

FPGA: Field-Programmable Gate Array. An integrated circuit that can be programmed by the user and that provides multilevel logic.

FUZZIFICATION: The process of converting a crisp number or set to a fuzzy number or set.

FUZZY CONTROL: A control system in which the controller is based on an algorithm composed of IF/THEN rules. Since more than one rule may fire (the conditions for their applicability occurs) with varied strength at the same time, a defuzzification process follows to generate a crisp control action.

FUZZY C-MEANS: A data clustering technique in which each data point belongs to a cluster to a degree specified by a membership function.

FUZZY ENTROPY: A measure of the fuzziness of a set. The more a set resembles its negation, the greater its fuzzy entropy, and the fuzzier it is.

FUZZY INFERENCE: The process of mapping an input space to an output space using fuzzy reasoning. Various types of inference processing have been suggested and used, including:

Mamdani-type inference in which the fuzzy sets from the consequent of each rule are combined through the aggregation operator and the resulting fuzzy set is defuzzified to yield the output of the system.

Sugeno-type inference (sometimes referred to as Takagi-Sugeno-Kang, TSK, method) in which the consequent of each rule is a linear combination of the inputs. The output is a weighted linear combination of the consequents. It does not involve a defuzzification process.

FUZZY LOGIC: A scheme of systematic analysis that uses linguistic variables, such as hot, cold, very, little, large, small, etc., as opposed to Boolean or binary logic, which is restricted to true or false states. The objective of the scheme is to enable the computer to make human-like decisions.

FUZZY MODIFIER: An added description of a fuzzy set, such as *very, very very*, that leads to an operation that changes the shape (mainly the width and position) of a membership function.

FUZZY OPERATORS: Operators that combine fuzzy antecedents to produce truth values. Zadeh defined fuzzy operators in terms of min-max rules. Several other methods of definition exist: the alternatively defined operators are referred to as non-Zadeh operators or compensatory operators.

FUZZY SET: A set that allows elements to have a degree of membership ranging from one to zero. The objective is to define precisely what is intrinsically vague. A crisp set allows only two values of membership, either one or zero. An ultra-fuzzy set has its membership function itself as a fuzzy set.

GENERALIZATION: In neural networks, it is the ability of the network to respond correctly to new input patterns to which the network was not trained, e.g. corrupted patterns. The ability to generalize a feature that distinguishes training from programming a system.

GÖDEL'S THEOREM: The crucial result of Gödel's theorem (after Kurt Gödel (1906-78)) is that within any given branch of mathematics there would always be some propositions that couldn't be proven either true or false using the rules and axioms of that mathematical branch itself. The implication is that all logical systems of any complexity are incomplete; each of them contains more true statements than it can possibly prove according to its own defining set of rules. An external audit is required! It also implies that the logicist doctrine of the deductibility of all mathematics from the axioms of logic is false. Gödel's Theorem has been used to suggest that no computational machine can ever be as intelligent as humans because the extent of its knowledge is limited by a fixed set of axioms. Humans, contrary to the computational machines *we envision* now, can discover truths previously unknown to them. The theorem has been taken also to imply that one will never entirely understand oneself, since one's brain can only be sure of what it knows about itself by relying on what it knows about itself.

HARDWARE\SOFTWARE CO-DESIGN: The simultaneous design of hardware and software components to meet system requirements.

HARVARD ARCHITECTURE: A processor architecture in which program and data storage address spaces are separate.

HEDGES: Linguistic terms that intensify, dilute, or complement a fuzzy set.

HOPFIELD NETWORK: A single-layer that consists of a number of neurons (nodes); each is connected to every other neuron. The weights on the link from one neuron to another are the same in both directions. The network takes only two-state inputs.

HYPER- A prefix to indicate that the geometrical form to follow is multi-dimensional.

HYPER-CUBE: A cube in an n-dimensional space. A crisp set defines a corner of a unit hyper-cube, while a fuzzy set defines a point inside the hyper-cube.

Isotopic Engineering: A concept put forward by A. A. Berezin in the 1980's that refers to purposefully redistributing the stable isotopes occurring naturally in materials for information storage and processing.

JIT Compiler: Just-In-Time compiler. Compiles program sections on demand during execution.

Kohonen Network: A self-organizing neural network. All the neurons are in one two-dimensional layer or grid with inputs connected to every neuron. Every neuron is connected laterally to its immediate neighbours.

Latin Phrases & Abbreviations:

i.e. (id est): that is, in other words.
e.g. (exampli gratia): for example.
viz. (videlicet): namely
vs. (versus): against.
etc. (et cetera): and so forth.
et al. (et alii): and others.
N. B. (nota bene): note well.
cf. (conferre): compare.
q.v. (quod vide): see reference.
QED (quod erat demonstrandum): that which was to be proven.
ad hoc: arranged for this purpose.
ad infinitum: to infinity.
a posteriori: from effect to cause.
a priori: already known.
ab initio: from the beginning.
de facto: in reality.
ergo: therefore.
in situ: in its original place.
modus operandi: a characteristic method of work.
per se: in and by itself.
verbatim: word-for-word.

Learning Law: The rule used during training of a neural network for systematic updating of their weights.

Linguistic Variable: Common language expression used to describe a condition or a situation, such as *hot, cold*, etc. It can be expressed using a fuzzy set defined by the designer.

Logic: The investigation (and application) of correct and reliable reasoning and inference.

MANY-VALUED LOGIC: Many-valued logics are non-classical logics. They are similar to classical logic in the sense that they employ the principle of truth functionality. They differ from classical logic by allowing truth degrees. Fuzzy logic is in a way a many-valued logic that, through the use of linguistic modeling, led to industrial and business applications.

MEMBERSHIP FUNCTION: The mapping that associates each element in a set with its degree of membership. It can be expressed as discrete values or as a continuous function. Triangular and trapezoidal membership functions are commonly used for defining continuous membership functions.

MICROCONTROLLER: A microprocessor designed specifically for embedded systems application. Microcontrollers typically include a CPU (central processing unit), memory, and other peripherals on the same chip.

MOORE'S LAW: The gist of the so-called Moore's law is that the number of transistors in integrated circuits increases many times every year. There are various versions and amendments to the actual statement put forward over the years. The original statement was an observation made in 1965 in an article for *Electronics* magazine by Gordon Moore to the effect that the number of the transistors per square inch had doubled since the integrated circuit was invented. He predicted that the trend would continue for the next ten years. (At the time, he was the Head of R&D at Fairchild Semiconductor Corp and in 1968 he and Robert Noyce founded Intel, then he became the Chair Emeritus since 1997.) In 1975, Gordon Moore amended the statement to say that the doubling would occur every two years. The present version is that the *data density doubles every 18 months*. The statement has also been expressed as: (bits per square inch) $= 2^{(t-1962)}$, where t is time in years. Some view Moore's law as a self-fulfilling prophecy. It puts pressure on semiconductor manufactures to strive to meet the predictions for fear that the competition would do that first. One can think of it as a statement of mission rather than a law of nature.

Increased density of transistors leads to increased processing speed, but also results in increased power consumption and hence the generation of excessive amounts of heat. Manufacturers may have to change circuit design to reduce the standard voltage used, a task that is not simple.

A survey of the IEEE Fellow Members on Technology Trends that appeared in the January 2003 issue of the IEEE Spectrum included the question: *How long do you think Moore's law (doubling integrated circuit transistor density every 18 months) will continue to hold true?* About 52% of the respondents selected 5-10 years, about 31% selected less than 5 years, and the rest selected more than 10 years. Since no technical reasons were given for such predictions nothing is confirmed, one can only muse at the results as is the case in many surveys of similar nature. They, however, confirm and expand on Ernest Rutherford's observation that *the only possible conclusion the social sciences can draw is: some do, some don't.*

There are, of course, physical limits on how small a transistor structure can be without having to deal with quantum effects. Joel Birnhaum (Hewlett-Packard senior vice president of R&D) was quoted to say: *in 2010 we will run into the physical limitation of having a fraction of an electron show up at the gate to switch the state of the transistor.* Nevertheless, it has been argued that there are ways to increase the number of transistors without reaching physical limits. Employing wider chips is one possibility, thicker chips is another. On should expect, or even hope for our own progress, that fundamentally new technologies will emerge changing the nature of computing and rendering Moore's law irrelevant, just a historic curiosity. One should remember that perfecting the fabrication of the vacuum tube was interrupted by the invention of the transistor. The law would survive longer if amended further to imply the doubling (or n-tupling, with the value of n to be determined) of the computing device. Some believe that it is economy, not physics or technology, that will invalidate Moore's law in any of its forms. The shrinking return on investment will reduce the commercial appeal of technological advances.

NANO-TECHNOLOGY: Technology development at the atomic, molecular or macromolecular levels, in the length scale of approximately 1-100 nanometer range to create and use structures, devices and systems that have novel properties and functions.

NEURON: The biological neuron is a basic cell of the nervous system. The dendrites function as its inputs, the axon functions as its output, and the synapse is the junction between the axon of one neuron and the dendrite of another. When a neuron is stimulated at the dendrite, it sums up the incoming potentials. If the sum is high enough, it sends an action potential down the axon. The operation is modeled in an electronic neural network by a processing element, PE, that performs a weighted summation and has a threshold and a sigmoid-type transfer function. In an artificial neural network the processing unit that models the biological neuron may be referred to as a node, neurode, or simply a neuron if no confusion with the biological one is feared.

OBJECT CODE: A program in binary code.

PERCEPTRON: A single-layer neural network that performed the first training algorithm. It can solve only linearly separable problems.

PID CONTOLLER: A cascade control device inserted in the forward path of a feedback control system. Its input is the error signal and its output is the control action. It consists of three control terms: proportional, integral, and derivative. Increasing the gain of the proportional term increases the speed of the system's response and *reduces* the steady state error, but tends to destabilize the system. The integral term can *remove* the steady-state error, but tends to destabilize the system because of the extra phase lag it introduces. The derivative term speeds up the transient response; it introduces a phase-lead and thus has a stabilizing effect.

PROCESSOR-INDEPENDENT: A piece of software that is independent of the processor on which it will run. Software written in a high level-language such as C/C++ can be made largely processor-independent, by rigorous design for portability while software written in assembly language is processor-dependent. Java is an example of a truly processor-independent language.

RAM: Random Access Memory. A broad classification of memory devices that includes all devices in which individual memory locations are accessible for reading and writing.

RISC: Reduced Instruction Set Computer. It describes the architecture of a microprocessor family that generally features fixed-length op-codes, a load-store memory architecture, and a large number of general-purpose registers.

ROM: Read Only Memory. A memory with fixed contents.

QUANTUM COMPUTING: A suggested mode of computing that is based on the principles of quantum physics which allow a particle, e.g., an electron, to exist in multiple states at the same time. The two most relevant aspects of quantum physics are the principles of superposition and entanglement. If quantum computing becomes practical, computing performance would achieve billion-fold gains. Centers of research in this area include: Oxford University, IBM, Los Alamos National Laboratory, and MIT.

SIMULATED ANNEALING: The process of introducing random noise to the weights and inputs of a neural network then gradually reducing it. The concept is similar to that of metallurgical annealing where the system starts with a high temperature to avoid local energy minima, and then gradually the temperature is lowered according to a particular algorithm.

SINGLETON: A set that has only one member.

SMART MATERIALS & STRUCTURES: Smart, intelligent, active, or adaptive materials are terms sometimes used interchangeably. They refer to materials that can sense external stimuli and respond with active control to those stimuli in real or near-real time. Smart materials are characterized by abrupt change once the material reaches its transformation temperature. The so-called shape memory alloy has the ability to revert to its trained shape when heated to its transformation temperature, the ability to be trained several times to different shapes and the ability to recover completely from a plastic deformation. Smart structural systems rely on embedded functions of sensors, actuators and processors, that can automatically adjust structural characteristics in response to the change in external disturbance and environments, thus giving the structure a degree of autonomy. Smart structure actuators include shape memory alloys, piezoelectric and electrostrictive ceramics, magnetostrictive materials, and electro- and magneto-rheological fluids and elastomers. Combining the concepts of smart structures, neural networks, and fuzzy logic could lead to powerful applications.

SOFT COMPUTING: Computing that is tolerant of imprecision, uncertainty, partial truth, and approximation. Its role model is the human mind. It includes fuzzy logic, neural computing, evolutionary computing, machine learning, and probabilistic reasoning.

SOS: System-On-Silicon. A single-chip system that includes computation, memory, and I/O.

STABILITY-PLASTICITY PROBLEM: The situation when a neural network is not able to learn new information without the destruction of previous learning; a time-consuming retraining will be needed every time new information is to be learned. ART networks solved this problem.

T-CONORM: It is also known as **S-NORM**. It is a two-input function that describes a superset of fuzzy union operators, including maximum, and algebraic sum.

T-NORM: A two-input function that describes a superset of fuzzy intersection operators, including minimum, and algebraic product.

TRUTH FUNCTIONALITY: The principle that the truth of a compound sentence is determined by the truth values of its component sentences. Thus, it remains unaffected when one of its component sentences is replaced by another that has the same truth value.

TURING'S TEST: The Turing test (after Alan Turing (1912–54)) consists of a human being, A, with a typewriter-like or TV-like terminal with which A can communicate with two sources unknown to him: B and C. One source is controlled by a machine and the other by a human unknown to A. A has to determine which of B and C is controlled by the machine and which is controlled by the human. If A cannot distinguish one from the other with accuracy significantly higher than 50%, and if the results persist regardless of what individuals are involved in the experiment, then one can say that the machine has *simulated* human intelligence.

VON NEUMANN ARCHITECTURE: A processor architecture that stores instructions and data in the same address space.

Index

ELSEVIER SCIENCE CD-ROM LICENSE AGREEMENT

of the Proprietary Material that are the results of discrete searches; (b) alter, modify, or adapt the CD-ROM Product, including but not limited to decompiling, disassembling, reverse engineering, or creating derivative works, without the prior written approval of Elsevier Science; (c) sell, license or otherwise distribute to third parties the CD-ROM Product or any part or parts thereof; or (d) alter, remove, obscure or obstruct the display of any copyright, trademark or other proprietary notice on or in the CD-ROM Product or on any printout or download of portions of the Proprietary Materials.

RESTRICTIONS ON TRANSFER

This License is personal to You, and neither Your rights hereunder nor the tangible embodiments of this CD-ROM Product, including without limitation the Proprietary Material, may be sold, assigned, transferred or sub-licensed to any other person, including without limitation by operation of law, without the prior written consent of Elsevier Science. Any purported sale, assignment, transfer or sublicense without the prior written consent of Elsevier Science will be void and will automatically terminate the License granted hereunder.

TERM

This Agreement will remain in effect until terminated pursuant to the terms of this Agreement. You may terminate this Agreement at any time by removing from Your system and destroying the CD-ROM Product. Unauthorized copying of the CD-ROM Product, including without limitation, the Proprietary Material and documentation, or otherwise failing to comply with the terms and conditions of this Agreement shall result in automatic termination of this license and will make available to Elsevier Science legal remedies. Upon termination of this Agreement, the license granted herein will terminate and You must immediately destroy the CD-ROM Product and accompanying documentation. All provisions relating to proprietary rights shall survive termination of this Agreement.

LIMITED WARRANTY AND LIMITATION OF LIABILITY

NEITHER ELSEVIER SCIENCE NOR ITS LICENSORS REPRESENT OR WARRANT THAT THE INFORMATION CONTAINED IN THE PROPRIETARY MATERIALS IS COMPLETE OR FREE FROM ERROR, AND NEITHER ASSUMES, AND BOTH EXPRESSLY DISCLAIM, ANY LIABILITY TO ANY PERSON FOR ANY LOSS OR DAMAGE CAUSED BY ERRORS OR OMISSIONS IN THE PROPRIETARY MATERIAL, WHETHER SUCH ERRORS OR OMISSIONS RESULT FROM NEGLIGENCE, ACCIDENT, OR ANY OTHER CAUSE. IN ADDITION, NEITHER ELSEVIER SCIENCE NOR ITS LICENSORS MAKE ANY REPRESENTATIONS OR WARRANTIES, EITHER EXPRESS OR IMPLIED, REGARDING THE PERFORMANCE OF YOUR NETWORK OR COMPUTER SYSTEM WHEN USED IN CONJUNCTION WITH THE CD-ROM PRODUCT.

If this CD-ROM Product is defective, Elsevier Science will replace it at no charge if the defective CD-ROM Product is returned to Elsevier Science within sixty (60) days (or the greatest period allowable by applicable law) from the date of shipment.

Elsevier Science warrants that the software embodied in this CD-ROM Product will perform in substantial compliance with the documentation supplied in this CD-ROM Product. If You report significant defect in performance in writing to Elsevier Science, and Elsevier Science is not able to correct same within sixty (60) days after its receipt of Your notification, You may return this CD-ROM Product, including all copies and documentation, to Elsevier Science and Elsevier Science will refund Your money.

YOU UNDERSTAND THAT, EXCEPT FOR THE 60-DAY LIMITED WARRANTY RECITED ABOVE, ELSEVIER SCIENCE, ITS AFFILIATES, LICENSORS, SUPPLIERS AND AGENTS, MAKE NO WARRANTIES, EXPRESSED OR IMPLIED, WITH RESPECT TO THE CD-ROM PRODUCT, INCLUDING,

WITHOUT LIMITATION THE PROPRIETARY MATERIAL, AN SPECIFICALLY DISCLAIM ANY WARRANTY OF MERCHANTABILITY OR FITNESS FOR A PARTICULAR PURPOSE.

If the information provided on this CD-ROM contains medical or health sciences information, it is intended for professional use within the medical field. Information about medical treatment or drug dosages is intended strictly for professional use, and because of rapid advances in the medical sciences, independent verification of diagnosis and drug dosages should be made.

IN NO EVENT WILL ELSEVIER SCIENCE, ITS AFFILIATES, LICENSORS, SUPPLIERS OR AGENTS, BE LIABLE TO YOU FOR ANY DAMAGES, INCLUDING, WITHOUT LIMITATION, ANY LOST PROFITS, LOST SAVINGS OR OTHER INCIDENTAL OR CONSEQUENTIAL DAMAGES, ARISING OUT OF YOUR USE OR INABILITY TO USE THE CD-ROM PRODUCT REGARDLESS OF WHETHER SUCH DAMAGES ARE FORESEEABLE OR WHETHER SUCH DAMAGES ARE DEEMED TO RESULT FROM THE FAILURE OR INADEQUACY OF ANY EXCLUSIVE OR OTHER REMEDY.

U.S. GOVERNMENT RESTRICTED RIGHTS

The CD-ROM Product and documentation are provided with restricted rights. Use, duplication or disclosure by the U.S. Government is subject to restrictions as set forth in subparagraphs (a) through (d) of the Commercial Computer Restricted Rights clause at FAR 52.22719 or in subparagraph (c)(1)(ii) of the Rights in Technical Data and Computer Software clause at DFARS 252.2277013, or at 252.2117015, as applicable. Contractor/Manufacturer is Elsevier Science Inc., 655 Avenue of the Americas, New York, NY 10010-5107 USA.

GOVERNING LAW

This Agreement shall be governed by the laws of the State of New York, USA. In any dispute arising out of this Agreement, you and Elsevier Science each consent to the exclusive personal jurisdiction and venue in the state and federal courts within New York County, New York, USA.